Power Devices for Efficient Energy Conversion

Power Devices for Efficient Energy Conversion

Gourab Majumdar
Ikunori Takata

PAN STANFORD PUBLISHING

Published by

Pan Stanford Publishing Pte. Ltd.
Penthouse Level, Suntec Tower 3
8 Temasek Boulevard
Singapore 038988

Email: editorial@panstanford.com
Web: www.panstanford.com

British Library Cataloguing-in-Publication Data
A catalogue record for this book is available from the British Library.

Power Devices for Efficient Energy Conversion

Copyright © 2018 Pan Stanford Publishing Pte. Ltd.

All rights reserved. This book, or parts thereof, may not be reproduced in any form or by any means, electronic or mechanical, including photocopying, recording or any information storage and retrieval system now known or to be invented, without written permission from the publisher.

For photocopying of material in this volume, please pay a copying fee through the Copyright Clearance Center, Inc., 222 Rosewood Drive, Danvers, MA 01923, USA. In this case permission to photocopy is not required from the publisher.

ISBN 978-981-4774-18-5 (Hardback)
ISBN 978-1-351-26232-3 (eBook)

Contents

Preface	ix

1 Introduction — 1

1.1	Era Predating the Birth of Power Semiconductor Devices	2
1.2	Dawn of Power Electronics	5
1.3	Japan's Leading Effort in Power Electronics	7
1.4	Chronology of Power Devices in the 1980s	10
1.5	Chronology of Power Modules in the 1980s	11
1.6	History of Recent Power Switching Semiconductors	12
1.7	Key Role of Power Devices for Efficient Power Conversion	15
	1.7.1 Role of Devices in Power Amplification	15
	1.7.2 Role of Devices in Power Switching Applications	21
	1.7.2.1 Power losses by power switches in power conversion electronics	23

2 Basic Technologies of Major Power Devices — 27

2.1	Power Device Categories	27
2.2	Key Semiconductor Operation Principles	28
	2.2.1 Essence of the Power Device	28
	2.2.2 Characteristics of the Semiconductor	32
	2.2.3 p-Type and n-Type Semiconductors	36
	2.2.4 Potential Barrier between Regions Having Different Impurity Concentrations	39
2.3	Basic Operation of Power Devices	43
	2.3.1 Reverse Voltage Blocking	43
	2.3.2 Forward Conducting	51

| | | 2.3.3 | Voltage-Holding Ability with Large Current: SOA | 54 |

2.3.3 Voltage-Holding Ability with Large Current: SOA — 54

2.4 Diode Rectifiers — 60

2.4.1 Diode Structures — 60

2.4.2 Transient Operation of a pin Diode — 65

2.4.3 Basic Operation of a pin Diode — 72

2.4.4 High-Voltage Large-Current Operation of a pin Diode — 81

2.5 Fast-Recovery Diode for a Typical Freewheeling Function — 83

2.5.1 Need for First-Recovery Diodes — 83

2.5.2 Effect of Lifetime Control — 87

2.5.3 Control Methods of the Lifetime — 90

2.5.4 Various Recombination Models — 93

2.5.5 Leakage Current Caused by Lifetime Killers — 99

2.5.5.1 Pair generation leakage current — 99

2.5.5.2 Diffusion leakage current — 101

2.5.6 Interpretation of Observed J_F–V_F Characteristics — 103

2.6 Devices of the Thyristor Family — 105

2.6.1 Thyristor — 105

2.6.2 GTO/GCT — 107

2.7 Bipolar Junction Transistors — 112

2.7.1 BJT Structures — 112

2.7.2 Basic Operation of the BJT — 124

2.7.3 High-Voltage Large-Current Operation of the BJT — 129

2.7.4 Safe Operating Area of the BJT — 135

2.8 Metal-Oxide-Semiconductor Field-Effect Transistors — 136

2.9 Insulated Gate Bipolar Transistors — 144

2.9.1 IGBT Structures — 144

2.9.2 Basic Operation of the IGBT — 158

2.9.3 High-Voltage Large-Current Operation of the IGBT — 166

2.9.4 Safe Operating Area of the IGBT — 170

2.9.4.1 Observation of IGBT destructions — 171

2.9.4.2 Destruction mechanism of real IGBTs — 175

3 Applied Power Device Family: Power Modules and Intelligent Power Modules 181

3.1 Review of the Power Module Concept and Evolution History 181

3.2 Power Module Constructional Features and Design Aspects 185

3.2.1 Basic Aspects of Power Module Construction and Design 189

3.2.1.1 What are the characteristics required from a power module package? 189

3.2.1.2 What are the features and issues related to typical power module package designs? 192

3.2.2 Fundamentals of Power Module Structural Reliability and Life Endurance 197

3.3 State-of-the-Art Key Power Module Components 207

3.3.1 Dual-in-Line Intelligent Power Module 208

3.3.2 Intelligent Power Module 211

3.3.2.1 A review of the IPM's fundamental concept 213

3.3.2.2 Chip technologies driven by the IPM evolution 217

3.4 Tips on and Guidelines for Applying Power Modules 236

3.4.1 Formulas Common for Power Device Driver Circuit Designs 236

3.4.2 Structure and Operation of the IGBT Device Used in a Power Module 237

3.4.2.1 Review of power MOSFET and IGBT basic operation principle 240

3.4.2.2 Review of a parasitic thyristor's destructive latch-up in an IGBT cell 248

3.4.3 Power Circuit Design 248

3.4.3.1 Turn-off surge voltage 249

3.4.3.2 Freewheeling diode recovery surge 250

3.4.3.3 Design issues related to ground loops 253

3.4.3.4 Reducing gate circuit inductance and avoiding mal-triggering 256

viii | Contents

	3.4.3.5	Power circuit impedance and overvoltage protection	260
3.4.4	Power Circuit Thermal Design Aspects		265
	3.4.4.1	Estimating power losses	266
	3.4.4.2	Estimating the junction temperature	271

4 Future Prospects — **273**

4.1 Summarizing Device Achievements — 274

4.2 Future Prospects of Silicon-Based Power Device Technologies — 278

    4.2.1 Overview of the Power MOSFET and the Superjunction MOSFET — 278

        4.2.1.1 Power MOSFET — 278

        4.2.1.2 Superjunction MOSFET — 281

    4.2.2 Review of IGBT Chip Technology and Its Future Prospects — 283

    4.2.3 Review of Diode Chip Technology and Its Future Prospects — 286

    4.2.4 Integrated Power Chips Combining IGBT and Diode Functions — 291

        4.2.4.1 Reverse-conducting IGBT — 292

        4.2.4.2 Reverse-blocking IGBT — 293

4.3 Prospect of Using Wide-Bandgap Materials — 295

    4.3.1 WBG Material–Based Advanced Power Chip Technologies — 296

    4.3.2 Benefits from SiC Application — 299

    4.3.3 Status of SiC Devices — 301

    4.3.4 Looking at Application Ranges for WBG Devices — 313

Bibliography — 317

Index — 323

Preface

Recently the role of power electronics and power devices in addressing the challenges in power and energy conversion and storage has been given wide attention as climate change has become a crucial global issue. In power electronics applications, the power density factor related to system design has improved remarkably in the past two decades. The main contributions in this growth have come from timely development of newer power modules, achieved through multidimensional major breakthroughs in insulated gate bipolar transistor (IGBT) and other power chip technologies; packaging structures; and functionality integration concepts. Driven by various application needs in the past decades, various generations of the IGBT and intelligent power modules (IPMs) have evolved so far and have been widely applied in different power electronics devices, covering industrial motor controls, household appliances, railway traction, automotive power-train electronics, windmill and solar power generation systems, etc. Additionally, aggressive R&D achievements have been made in wide-bandgap (WBG) power devices, especially by employing SiC as the base semiconductor material.

In this book, state-of-the-art power device technologies, their physics and operating principles, features related to design, and tips for applications that have driven the various device evolutions will be discussed. Also, the major development trends in the areas of power chip and module technologies will be introduced, including perspectives of WBG devices.

G. Majumdar
I. Takata

Chapter 1

Introduction

Several evolutionary changes and breakthroughs achieved in the areas of power semiconductor device physics and process technologies have accelerated the growth of power electronics in recent years, centering on inverters and converters as its key system topologies.

Actually, the view is that the electronics revolution caused by semiconductor devices started in the signal and information processing area and spread to the power conversion and control area in the 1960s. The appearance of the thyristor (or silicon-controlled rectifier [SCR]) in 1957 as a semiconductor device with excellent power-handling capability, like a solid-state relay/switch, brought revolutionary progress in electric power conversion and established a new technology known as "power electronics." In this chapter a brief review of the history of power devices will be made, followed by discussions on the necessity of expanding usage of power electronics to support our energy conscious social environment and on some of the key roles that power devices play in this respect for efficient power conversion.

Power Devices for Efficient Energy Conversion
Gourab Majumdar and Ikunori Takata
Copyright © 2018 Pan Stanford Publishing Pte. Ltd.
ISBN 978-981-4774-18-5 (Hardcover), 978-1-351-26232-3 (eBook)
www.panstanford.com

1.1 Era Predating the Birth of Power Semiconductor Devices

Active electrical devices have a history dating back to 1906, starting with the invention of the triode. History says that these triodes could be replaced by a much smarter active device solution, named "transistor," after its invention in 1947. The germanium grown transistor type was commercialized in 1953. Prior to the invention of the triode, the need for electrical devices could also be traced back to the late second half of the nineteenth century. For instance, in the 1880s, tailoring of alternating current in power transmission required electrical devices. Also, large current alternating current (AC) to direct current (DC) conversion was needed in the electrolysis and electrolytic refining processes and so on. For these applications a high-current DC generator was said to have been developed in the 1880s. Basically it was a large-sized system comprising an AC motor designed to turn a DC generator producing high current at low voltage.[1] Additionally, in 1881, the Ward Leonard system shown in Fig. 1.1 was invented to control precisely a DC motor through an induction motor and a DC generator. This method is still being used by servicing industries for maintenance and servicing work on old elevators in different regions even today.

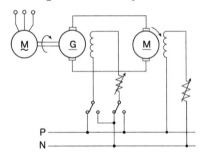

Figure 1.1 The Ward Leonard system.

After the introduction of the triode concept, a new modified version of the discharge tube, named "thyratron," was introduced around 1910. A thyratron enclosed a small amount of gas (mercury,

---
[1] For instance, 6 V/600 A DC generators were manufactured in the 1880s.

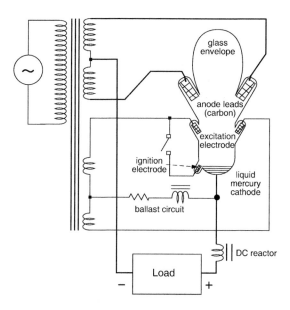

Figure 1.2 A DC supply circuit using a mercury rectifier.

xenon, hydrogen, etc.) in a triode discharge tube, aiming for sensitivity improvement of the original vacuum tube. However, an operating mode was discovered, which could continue turning on electricity as long as voltage was applied, by forming plasma with electrons and ions separated from the gas in the whole discharge tube. This kind of operation could conduct a far superior electric current in comparison with the vacuum tube, which only had a mechanism of thermoelectron run-through.

Thyratron, which used mercury for its cathode, did not necessarily require heating of its cathode, unlike those methods that used other gases, and could also treat high current. Therefore, it was also called a mercury rectifier (Fig. 1.2).[2] The device was able to control a conducting point using an excitation electrode, but the handling was considerably complicated. The most serious problem was malfunctioning caused by the device's tendency to turn on abnormally without any signal. In spite of such issues, thyratron

---

[2] The ignition electrode strikes an arc at the start, and the excitation electrode keeps the arc.

products of 2 kV/8 kA class appeared around 1930. Thyratrons were adopted mainly in high-voltage circuits because their on-voltage was very high (around 10 V).

In 1955 Japanese National Railways fabricated an experimental AC drive locomotive "ED45" using three DC motors of total 1 MW output capacity driven by four ignitrons, which were constructed by a single-phase bridge rectifier on the secondary low tap-voltage output terminals of a 20 kV AC transformer. The ignitron used was a water-cooled single-phase mercury rectifier having an improved ignition method, and it needed a mercury-vapor diffusion pump and a rotary pump to maintain a specific vacuum level in the iron container. Then in the world's first universal locomotive "EF30" series, which were made starting from 1960, the single-ignitron system was replaced by twenty 700 V/200 A silicon (Si) diodes connected in a 10-in-series and 2-in-parallel circuit architecture. The very first series, Shinkansen bullet trains, which came up in 1964 used a control system as that of "EF30"[3] except that the overhead wire supply voltage was as high as 25 kV. In the Sakuma frequency conversion facility (300 MW) that operated from 1965, 26 mercury rectifiers of 125 kV/1.2 kA were used until 1993. A real breakthrough in the solid-state semiconductor-based three-terminal-based device was when the General Electric (GE) Company, of the US, successfully introduced a 300 V/7 A device named SCR in 1957. As this SCR performed the same operation as of a thyratron, it was named "thyristor" after its predecessor, thyratron, in the 1963 International Electro-technical Commission. Prior to thyristor introduction, the epoch making "transistor," which W. Shockley invented in 1947, existed as a three-terminal solid-state semiconductor concept device. However, this solution was inadequate for high-voltage/current conversion requirement in the early days. The transistor technology shifted to silicon-material-based device fabrication technology from the original germanium-based approaches. However, due to shortcomings of the germanium semiconductor material (difficult-to-make devices of voltage breakdown characteristics higher than 300 V and a

---

[3]The group having 20 diodes was replaced by a five-series connection of 1300 V/ 300 A Si diodes.

temperature rating higher than 80°C),[4] thyristor and the diode technologies began with silicon-based fabrication methods, as the material was considered appropriate for making devices for high-power applications.

1.2 Dawn of Power Electronics

Presently, the term "electronics" generally attaches meaning closer to electronics in the fields of information and signal technologies. In this case, the contents of the transferred signal and processed data have major importance and energy and power have less meaning, as explained by Dr. E. Ohno in the book titled *Introduction to Power Electronics* [1]. Power electronics, which was newly defined by Dr. W. E. Newell in the early 1970s (explained also in the following sections), meant a new frontier of electronics application involving power and energy whereby all kinds of electronics related to the control of various electric machines, electric plants, and other power systems based on computer control and relays were meant to be covered [2]. More specifically, as also indicated by Dr. E. Ohno in his aforementioned book, power electronics implies technology dealing with the control of electric power in all its phases during generation, transmission, distribution, and conversion using power semiconductor devices such as thyristors, power diodes, and power transistors together with associated integrated circuits (ICs) and various other components. Therefore, the first specific feature of power electronics is its diversity in terms power and energy range it covers, from milliamperes/millivolts/milliwatts to several mega-amperes/megavolts/hundreds of megawatts. The other feature of power electronics is the extensive use power semiconductor devices, advancement of which had continuously sustained and complemented growth of this giant and very important electronics field.

---

[4]It was another reason that the selenium rectifier (30 V/50 mA/cm^2), which could handle much higher current than germanium, was put to practical use from about 1920. The selenium rectifier's limit temperature was the same level as germanium, but it was easy to make a large-area device and construct serial connections. Furthermore, it was much stronger in an overcurrent operation.

Thyristors and diodes using silicon as the power semiconductor material, explained later in this chapter, had progressed around a class of 3 kV/500 A device rating in about 1970 and triggered the start of solid-state semiconductor-based power electronics, replacing the bulky mercury rectifiers. In addition, as a drive means for DC motor control, the static Leonard control system, also called the "Thyristor Leonard system," had begun to be used in power conversion applications instead of the induction-motor-and-generator-combination-based classic "Ward-Leonard system" shown in Fig. 1.1.

Although the basic idea of using thyristors as controllable solid-state power semiconductor switches in power conversion existed from the 1960s, the term "power electronics" appeared much later. The name was submitted by a keynote address at the first Power Electronics Specialist Conference (PESC), in 1974, by Dr. Newell. He talked about the importance and growth possibilities of the technical area that should necessarily bind and extend to three fields—electronics, electricity, and control—and named it "power electronics." He also asked the participants for cooperation in that unripe new frontier. In addition, he predicted that power semiconductor switches would become indispensable components for determining the superiority and inferiority of applied power electronics topology and design in systems such as electric appliances and machine controls. Furthermore, he earnestly desired the development of an effective theoretical analysis technology and the appearance of a kind of "general-purpose switching device/module" that could turn off by itself. This desire motivated many development challenges in the decades of switching power semiconductor history but could only be truly satisfied by the invention and introduction of intelligent power modules (IPMs) by Mitsubishi Electric Corporation in the late 1980s, having silicon insulated gate bipolar transistors (IGBTs) as the core power switching element and dedicated self-diagnosis, protection, and control circuitry in an all-in-one module-structure-based concept.[5] The later chapters of this book will provide more details of IPM technology and its growth.

---

[5] An advanced power device integrating intelligent peripheral circuit functions along with core power switching transistors within a compact modular housing.

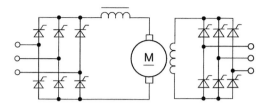

Figure 1.3 The static Leonard control system.

In the 1970s when Dr. Newell's appeal created a new sensation, the power devices' power switch application requirements were limited to the thyristor and diode devices. Although silicon transistors were realized, their operating capability in terms of power handling was very small and these devices were also not sufficient for high-voltage usage and were easily destroyable.

For this phenomenon, called "the second breakdown," special know-how was needed to drive an industrial motor with transistors. On the other hand, a gate turn-off thyristor (GTO) of around 1 kV/100 A, which, unlike an original thyristor, could perform a self-turn-off by a controlled gate current, had reached the state of trial production.

1.3 Japan's Leading Effort in Power Electronics

In the early 1980s, power electronics was finally becoming one field. And in 1983, Japan hosted the International Power Electronics Conference (IPEC) and showed enthusiasm for leading the world in power electronics technology. The drastic development of thyristors and power transistors in the latter half of the 1970s greatly contributed to this progress in Japan. In about 1980, the thyristor progressed to 3 kA 4 kV. In the GTO, a 2.5 kV 2 kA product was developed by Toshiba and was used at first for chopper control[6] to change the speed of the DC motor of the train. In those days, the

---
[6]A method to control DC voltage effectively by changing the ratio of the on/off period of the DC circuit.

8 | Introduction

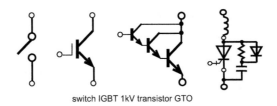

Figure 1.4 Switch architecture by the IGBT, the transistor, and the GTO.

Figure 1.5 Appearance and constitution of the early transistor module.

expectation from GTOs was high as a new device and seven to eight companies manufactured GTOs even in Japan.[7]

Around 1977, Fuji Electric manufactured a 450 V[8]/50 A Darlington transistor, which could be used for induction motor control or the welder of the AC 200 V line easily. A method to put a transistor together as shown in Fig. 1.4, which S. Darlington devised, was necessary for a high-voltage transistor because the higher-voltage transistor inevitably exhibits lower current gain ($h_{FE} \equiv I_C/I_B$).[9] The property of long-term high-voltage-withstanding endurance was particularly excellent because of introduction of planar technology,[10] which was the first case for a power device. In 1977 using such a chip, a 100 A transistor of the mold

---

[7] Now, all those companies, except one, have ceased production.

[8] It was customary in those days to express the withstanding voltage rating of the power transistor in V_{CEO}(sus). By the method of expressing in V_{CEX}, which became popular from about 1983 the device's rating denomination could have been changed to 600 V.

[9] Two-step constitution for 450 V and three-step constitution for 1000 V were necessary.

[10] It is the fundamental process in ICs, covering the surface of the chip with a silicon oxide film.

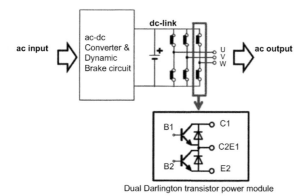

Figure 1.6 Voltage-type inverter architecture.

package with the emitter and base electrodes on the top surface was manufactured. And, in 1980, Mitsubishi Electric manufactured a 450 V transistor module having the following characteristics (Fig. 1.5 shows a 50 A × 2 module):

- Incorporation of two pairs of Darlington transistors and freewheeling diodes (FWDs).
- All terminals flattened, including the collector, for an easy bus bar connection.
- An electrically isolated heat dissipation base, which could be attached to a heat radiation fin directly.

When the general-purpose inverter applying three dual Darlington transistor power modules, as shown in Fig. 1.6, was first released in 1981, it drew a lot of attention because of its compact and handy features, along with its ability to control an induction motor's speed, transforming the rotating machine to a variable-speed-type device. The new general-purpose inverter's popularity also rose because of its moderate price tag, and soon the equipment got widely applied across the industry.[11] More than 20 Japanese companies commercialized it immediately. In 1982, these transistor modules had begun to be used for low-speed elevators. And home-use air conditioners using the inverter with six transistors and six FWDs

---

[11] The low price was possible on adapting a massive make-to-stock production.

in one type of module were also released by the companies. Strong heating and energy saving of the winter season were popular, and the name of the inverter spread across the standard homes in Japan.

In the same year, the 1000 V 300 A transistor module [3], which was available at AC 400 V line was sold and the general-purpose inverter began to spread rapidly in the industry. Until the beginning of the 1990s at least, transistor modules and GTOs made in Japan swept across the world. The power device for the inverter was divided into two areas. Transistor modules were used in the small and medium capacity, less than about 300 kW, such as high-speed elevators, and GTOs were used above it, in electric trains, the rolling mill at the steel manufacturer, and large power supplies.

With the various improvements of the control board using the LSIs and the downsized power capacitor, the contribution of the transistor module was indispensable for downsizing it and making the inverter lightweight. It omitted isolation with the heat sink and simplified the wiring using bus bars. In addition, as a result, it was able to reduce the stray inductance on the main circuit—a big contribution. With the improvement of the endurance capacity of the transistor chip, the module package made the transistor usable with the inverter.[12] It might be said that the transistor module was the first "general-purpose switching module" that Dr. Newell expected after all.

1.4 Chronology of Power Devices in the 1980s

The power transistor technology matured in 1982, and later development raised the current amplification factor h_{FE} for downsizing the base drive unit as much as possible. In 1982, GE produced IGBTs experimentally, calling them insulated gate rectifiers (IGRs). Toshiba succeeded in putting it to practical use and commercialized the IGBT module from 1985. It was the first half of the 1990s that the competitiveness of this IGBT module came to exceed that of the transistor module. Meanwhile, in the mid-1990s, the GTO

---

[12]The transistor module did not need the snubber, and a small capacitor between bus bars was used for surge protection.

reached 6 kV 6 kA and was completed as gate commutated turn-off thyristor (GCT) in 1996. Furthermore, in the late 1990s, the IGBT progressed and became large capacity and high voltage and 6 kV products were commercialized. And the IGBT module replaced the transistor module in not only industrial use but also home electric appliances. For the inverter ($\approx$1 kV) of the hybrid car that needs to be light weight and of small size, IPMs that contained IGBTs became indispensable. After about 1995, IGBTs and IPMs became mainstream in the railway vehicle, replacing the heavy and large GTO system. For instance, Shinkansen began to use GTOs of the 300 series in 1992. However, the new Shinkansen, such as 700 series, began using IGBTs from the end of 1998. In about 2000, IGBTs began to replace almost all GTO systems except the large electric capacity fields.[13] The golden years for IGBTs have been continuing since then. The road to realize effective switching power devices for power conversion applications, which Dr. Newell envisioned and demanded, may be said to have started getting engraved by the transistor power modules and finally fulfilled by realization of IGBT modules and IPMs.

1.5 Chronology of Power Modules in the 1980s

The package of the transistor module took an example from the thyristor module manufactured in Germany. However, long-term reliability of enduring high voltage was a problem because the transistor chip was of planar structure, which was vulnerable to adverse influences from the sealing material.[14] Furthermore, it had other weaknesses—use of an aluminum wire on the top surface and the current-crowding effect. In a heat cycle by the off and on operation, heat exhaustion due to the difference of the thermal expansion rate of different kinds of materials stood out. It was common sense to use the flat package, which sandwiched the chip by thick copper blocks and was used conventionally for thyristor

---

[13] GCTs have been selected for new systems in such fields.
[14] Because the thyristor chip attached a thick low-melting-point glass to the mesa structure part, it was hard for it to be affected by the outside world.

and GTOs in trains. However, the structure of the module greatly progressed, and as for the train, a module is mainstream now. But the flat package, which shorts out after destruction, is indispensable to the series connecting such an application of the DC transmission. In addition, a flat package is often used for the IGBT used in a large-capacity system more than several megawatts. By the way, although IGBTs have low driving power, high endurance capacity, and high-speed switching, a superior characteristic of the low on-voltage, various supporting circuits are necessary for an IGBT when one uses it, for instance, those for the protection for overcurrent, short circuit electric current, and abnormal temperature rise. Adding the driving circuit and its malfunction detection to these, IPMs, so to speak, aimed at realization of an "ideal switch." Because a main part of the inverter can be composed just to connect a power supply and a logic board to it, IPMs can be used across a wide range, from being downsized for household appliances to being used in cars.

1.6 History of Recent Power Switching Semiconductors

Figure 1.7 shows briefly the historical growth of power switching semiconductor devices. As explained in the previous part, in the 1960s the thyristor concept generated the first wave in the history of power switching semiconductor devices and opened up many possibilities for the growth of power electronics as a whole.

As highlighted, the bipolar transistor was invented in 1947 but the device was not structurally fit for use in sizable power conversion. Such insufficiency in solid-state device performance led to the invention of a pnpn device structure, called a thyristor (or a SCR), that was required to be latched on (thereby losing gate triggering controllability) for conducting current. However, by virtue of its capability in handling sufficiently high levels of current and voltage in power conversion topologies, the device became the prime choice for power electronics applications. In 1970, high-power thyristor devices were used in chopper circuitry for a railway traction application. In the second half of the 1970s the two controllable nonlatching type of devices, the bipolar transistor module and the

History of Recent Power Switching Semiconductors | 13

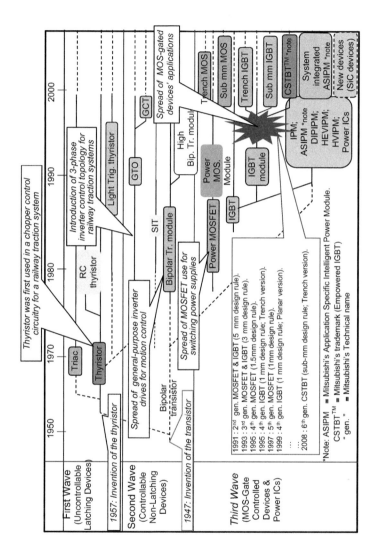

Figure 1.7 History of power devices.

GTO, were developed and introduced to match the growing demand for inverter-controlled power conversion equipment, which was generated as a new concept for saving energy in the early 1970s' "oil shock" and quickly became the focus of power electronics growth. This started the second wave in the chronological evolution of power semiconductor devices. The introduction of power metal-oxide-semiconductor field-effect transistors (MOSFETs) in the 1970s enabled compact and efficient system designs, particularly those based on low-voltage (less than 200 V) applications. In the beginning, the power MOSFET chips were fabricated using vertically diffused metal-oxide-semiconductor (VMOS) fabrication process. To improve performance and reliability, the double-diffused metal-oxide-semiconductor (DMOS) fabrication process technology was adopted, and this quickly became the predominant option for device manufacturers.

The thirst for performance improvement continued, and that drove device developers and manufacturers to commercially introduce trench gate MOSFETs by refining the basic structure and its manufacturing constraints in different ways. In the late 1980s through early 1990s the third wave started to build up, focusing on MOS-gated device physics blended with the bipolar transistor. As a

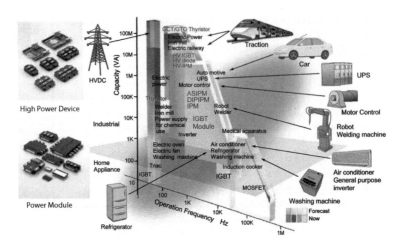

Figure 1.8 An application map of various power devices in terms of output capacity versus operating frequency of applied equipment/systems.

result, the evolutionary power device, the IGBT, was introduced and looked very promising as the key component for the acceleration of power electronics. Several technological efforts have made the IGBT positively comparable to the features of the bipolar transistor as well as to those of the power MOSFET. That is, the IGBT, by virtue of its MOSFET-like insulated gate controllability and bipolar-like conductivity-modulated on-state operation, has successfully shown its ability to perform adequately in high-power-and-frequency zones, as shown in Fig. 1.8.

1.7 Key Role of Power Devices for Efficient Power Conversion

Power devices can be defined to fall into two categories depending on their functions. One category includes power devices that operate as power switching elements in power electronic circuits. The other category includes devices that are used in high-frequency electronic circuits requiring high-linearity analog signal amplification operation.

1.7.1 Role of Devices in Power Amplification

Power amplification devices have a wide variety of applications. The basic role played by power devices in many of the fields is to amplify the power of a high-frequency input signal and to output a more powerful signal for further processing and usage. Because of their nonlinear characteristic, power amplification devices are also capable of operating to generate higher-frequency output signals from a specific input signal. However, here we will focus only on the pure power-amplification-related functionality of power devices for applications where the frequency of the signal at the input of the conversion amplifier remains unchanged at the output. The nonlinear characteristic of a device means that the relation between an input power (or voltage) and an output power does not have a straight-line relationship when plotted on a graph. A semiconductor diode is a simple example of a nonlinear power device as the current I flowing through it under a conducting state follows a nonlinear relationship with the voltage V across its terminals:

$$I = I_0 \exp^{V/V_T} \tag{1.1}$$

In high- and intermediate-frequency-range telecommunication systems amplifiers composed of power semiconductor transistors as their core components are in use either in discrete form or in IC form. The use of discrete components instead of the IC version depends on various factors, such as availability and the cost of systems versus performance targets. In terms of performance capability, the development of a new semiconductor process with higher transit frequency characteristics often makes use of materials or processes that are immature viewed from the productivity aspects. However, the high growth rate trend in the high-frequency communication systems in today's information & communication technology (ICT)–driven global environment has boosted the development of semiconductor processes that are apt for very-high-frequency applications. ICs based on compound semiconductor materials such as gallium arsenide (GaAs) or silicon germanium (SiGe) can be used to make transistors that scale a performance level up to the GHz range. In terms of device structure, bipolar transistors made using GaAs or SiGe are mainly used for application up to approximately 5 GHz level and are known as heterojunction bipolar transistors (HBTs). Above 5 GHz applications, GaAs material–based junction field-effect transistors (JFET) or metal-semiconductor field-effect transistors (MESFET) are commonly used as the key component in a power amplifier circuitry. As the transit frequency range is expanding from a few-GHz level to about a 100-GHz level along with higher-power-amplification and higher-conversion-efficiency requirements, newer compound semiconductor materials, such as gallium nitride (GaN), are being developed and gradually being applied for device manufacturing. Such requirement of higher-power-frequency domain emphasizes the need for a monolithically IC solution, that is, a semiconductor IC topology, incorporating smart, low-noise, highly efficient peripheral circuit integration rather than a discrete-component-assembled printed-circuit-board (PCB)–based design. To cope with such a trend, technologies related to the monolithic IC approach, called the monolithic microwave integrated circuit (MMIC), have become increasingly important.

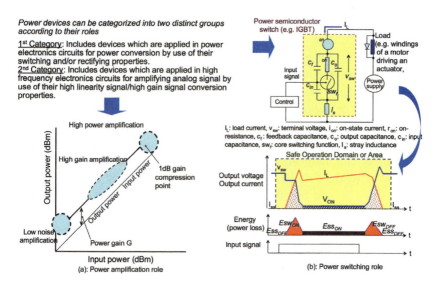

Figure 1.9 Roles of power devices (definition).

Power amplification performed by devices exhibit several features, such as high gain, low noise, high output, high efficiency, and wide bandwidth, depending on their applications.

Figure 1.9a explains the typical relationship in graphical form between the input signal power and the amplified output power of a typical power amplification circuit. Both input and output power signals are plotted typically on logarithmic scales in dBm values with 1 mW as the unit. Therefore, the amplification gain G directly becomes the deviated value from the "output power = input power" plot shown in the graph. Operation of power-amplification devices can be categorized into three basic forms on the basis of the power level of the applied input signal: low-noise amplification, high-gain amplification, and high-power amplification. In low-noise amplification, minimization of signal-to-noise ratio (S/N ratio) degradation under a very low signal power level environment becomes the key operation factor.

To meet such a need, devices of the low internal carrier dispersion effect type, such as a high-electron-mobility transistor (HEMT), made using a semiconductor material of the high carrier mobility type (typically, compound materials such as AlGaAs/GaAs or indium-implanted AlGaAs/InGaAS semiconductors), are preferable application choices. In such a HEMT device, an internal heterojunction is created on the compound material's high-mobility lattice face, forming the key active region of the transistor. An extremely high mobility type 2D electron gas, abbreviated as 2DEG, is formed below an activated gate region integrated between the drain and source regions of the transistor. A low-noise amplifier is an integral part of many systems, especially those used for wireless communications such as cell phones and wireless local area networks (WLANs). In such a system, typically a low-noise amplifier is used together with a preamplifier to boost the signal and reduce the noise, maintaining the S/N ratio required for the purpose. The power consumed by a low-noise amplifier and preamplifier assembly varies from application to application. In a typical consumer application such as a cell phone, such an assembly is designed to run on a low-voltage-biased supply network and consumes very low power while being designed to be of a very compact size. For a low-noise amplifier to be effective in boosting a signal without introducing much of its own noise, the first-stage amplification must be as high in gain as possible.

To facilitate the reception and transmission of signals in a low-power and high-noise environment, proper noise figure and power balancing are essential circuit design aspects. Stability of the frequency desired is another key design aspect. Furthermore, the input and output of the amplifiers are required to have good impedance matching to maximize their signal transfer functionality. This requires optimization of all components in terms of balancing the parameters of each variable involved. In high-gain amplification, power amplification becomes the priority. To achieve a high-power-gain operation the applied device should be able to operate at a very high oscillating frequency by virtue of its inherent high cut-off frequency and very low feedback capacitance characteristic features. For such purposes, bipolar junction transistors (BJTs) or field-effect transistors (FETs) structured using high-mobility semiconductor materials have been suitable device choices and advanced devices of the high-frequency linear-amplification-operation type have been continuously developed to fulfill various application needs. Additionally, having an insulated gate structural feature,

Key Role of Power Devices for Efficient Power Conversion | 19

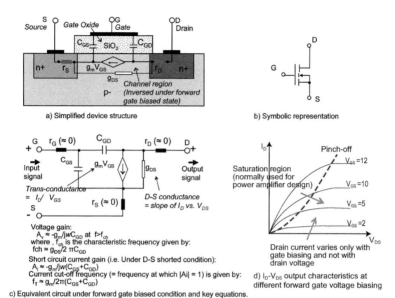

Figure 1.10 Lateral n-channel enhancement-mode MOSFET as a high-frequency amplifier.

MOSFETs are used in many applications as small-signal linear amplifiers. Figure 1.10 schematically describes a simplified device structure of a lateral n-channel enhancement-mode MOSFET device and its internal equivalent circuit, with the key equations and forward-gate-biased output characteristics showing a saturation region, which are relevant for the high-frequency-amplification function. For a transistor's high-frequency amplifier application, the voltage gain, short-circuit current gain, and cut-off frequency are expressed as follows (refer also to Fig. 1.10):

Voltage gain

$$A_v = \frac{V_O}{V_{GS}} \approx \frac{-g_m}{jwC_{GD}} \quad \text{at } f > f_{ch} \quad (1.2)$$

where f_{ch} is the characteristic frequency given by

$$f_{ch} \approx \frac{g_{DS}}{2\pi C_{GD}} \quad (1.3)$$

Short-circuit current gain (i.e., under $D - S$ shorted condition)

$$A_i \approx \frac{-g_m}{jw(C_{GS} + C_{GD})} \quad (1.4)$$

Current cut-off frequency (= frequency at which $|A_i| = 1$) is given by

$$f_T \approx \frac{g_m}{2\pi(C_{GS} + C_{GD})} \tag{1.5}$$

As the device is operated in the saturation region shown in Fig. 1.9c for power amplification, the gate-drain capacitance becomes very low due its variation behavior in relation with the drain potential with respect to the source. Also, in an actual circuit operating condition, the r_G, r_D, and r_S series resistances refer to Fig. 1.10c, which are approximated to be zero in deriving the characteristic frequency f_{ch} and the cut-off frequency f_T above for not being intrinsic features of the device, are to be taken into account in analyzing the maximum oscillation frequency f_{max} of the device. In other words, f_{max} is the characteristic of the device incorporating both extrinsic and intrinsic features and is defined as the frequency at which the power gain of the transistor device is equal to unity under a matched input-output impedance condition and can be expressed in a simplified form as follows:

$$f_{max} = \frac{f_T}{\sqrt{g_{DS}(r_G + r_D + r_S) + 2\pi r_G C_{GD} f_T}} \tag{1.6}$$

In the above expression, the resistances r_D and r_S are parasitic in nature, existing due to structural features. However, for a very small transistor design the values of these elements are very low and can be approximated to zero. Also, as the MOSFET is operated in its saturation region for high-frequency amplification, the value of g_{DS}, which is defined as the slope of I_D versus V_{DS} output characteristic refer to Fig. 1.10d, can also be approximated to be zero. Therefore, Eq. 1.6 can be simplified to define f_{max} as follows:

$$f_{max} = \sqrt{\frac{f_T}{2\pi r_G C_{GD}}} \tag{1.7}$$

For a high power-gain amplification design, transistors having a very high f_{max} feature are preferable device choices. Consequently, transistor devices that exhibit a high cut-off frequency f_T, together with low gate-drain feedback capacitance C_{GD} and low terminal gate resistance r_G features, are suitable for such applications.

1.7.2 Role of Devices in Power Switching Applications

Electric power is basically energy and in using electric power, the basic power-related characteristic elements, such as voltage, current, frequency, phase position, and number of phases, and these waveforms, which contain power-related information values, are converted, depending on the necessity for proper usage of power. For such conversion, an important goal is to make the process as efficient as possible by minimizing the associated power losses. To achieve such a goal, a new electronic technology to handle power using semiconductor devices as power switches in various circuit topologies was introduced in the 1960s and was termed "power electronics." In such circuits the power semiconductor switches perform controlled on-off actions, which become the basic response of the whole system [1]. And it obviously becomes essential that the semiconductor switches perform these actions continuously with minimized electric power losses. Power electronics can be broadly grouped into three subfields of technologies: power semiconductor device technology, power conversion circuit technology (nonlinear circuitry), and control technology used to operate the system using the former two technologies in combination. Figure 1.9b describes a single power switch circuit feeding a clamped inductive load from a power source. The power switch in this case can be chosen to be an n-channel IGBT or an npn bipolar transistor or an n-channel power MOSFET, depending on the circuit designer's needs. Each of on, on-state, off, and off-state actions of the power switch is controlled by the input command fed into the transistor generated by a switching algorithm given to the transistor's base or gate. The power switch and the associated circuit model to perform the required switching action can be very complicated depending on the topology used. Power conversion can be from AC to DC, DC to AC, AC to AC, or DC to DC output. The number phases for AC application, either on the input side or on the output side, can be a single-phase type or a three-phase type. Furthermore, for DC–DC conversion the output voltage can be an up-converter type or a down-converter type. And in AC–AC conversion, the frequency of transferring power through an isolation transformer or the frequency of the output AC voltage may require to be much higher than that of the voltage and/or current at the input circuit. The whole power conversion system thus may require a number of switching and nonswitching power devices, along with various other passive components and materials to make its structure, which can be as small as a tabletop compact unit or as large as a large-sized tank or truck. In all such applications, each power switching element is required to operate satisfying the following fundamental conditions, which are also reflected in the switching waveforms shown in Fig. 1.9b:

- The ability to perform switching with low switching loss (i.e., the low turn-on switching energy E_{sw} (ON) and low turn-off switching energy E_{sw} (OFF), described in Fig. 1.9b.)
- The ability to conduct current under a steady on-state with low conduction power loss (i.e., the low E_{ss} (ON) described in Fig. 1.9b.)
- The ability to operate under high-current and high-voltage conditions and exhibit high turn-off current capability retaining gate controllability (i.e., having appropriately large safe operating domain/area (SOA) in terms of voltage and current limits, as shown in Fig. 1.9b, and also in terms of time-withstanding ability under a high-current plasma state arising from an abnormal short-circuited condition.)
- The ability to stay in an off-state with low leakage current I_{ss}, reducing the off-state power loss (i.e., exhibiting a low E_{ss}, as described in Fig. 1.9b.)
- The ability to operate with low gate-circuit power loss (ideally the input signal drive is based on gate voltage–controlled device operation).

Thus, it can be said that the basic role of power devices is to perform on-off actions in power electronics circuits. It is worth mentioning here that contact or mechanical switches such as circuit breakers and relays are also popular elements often used for power conversion/handling circuits. These elements close and open electric circuits mechanically by induced electromagnetic force or induced pneumatic pressure or pneumatic pressure transmitted through a specially designed mechanical linkage. Due to contact resistance being very low (highly conductive metal-to-metal contact) the power loss of the entire power conversion operation is very low. However, the speed of operation of such elements being very low as well, and the life of the contact being limited, the application of these mechanical elements has mainly been for switching very high-power conversion ranges. In actual electric power systems, line switches and circuit breakers are the devices used to control the flow of electric power except for the excitation control of generators. In contrast to high-power electronic systems employing mechanical contact-based switches/circuit breakers, power electronics applications employ solid-state power switches as power conversion/flow control is achieved by high-frequency (>1 kHz) based switching algorithms for high efficiency and resolution. Referring to Fig. 1.9, it can be said that a high-power switching method draws far better advantage in terms of efficiency gain in power conversion, particularly in high-power-handling circuitry, compared to a linear amplification method. A linear amplification method as described in Fig. 1.9a can be the choice for power conversion or amplification when the output waveform needs to be high fidelity to its input. However, the conversion power losses generated in the process become very high, requiring a bulky and costly cooling method; consequently, the conversion efficiency becomes very low. Therefore, the linear amplification method has remained suitable only for applications with a very low power range (< several hundred watts), such as audio-visual amplifiers and low-power-range servo amplifiers. Now, we will examine the conditions for efficient power switching by power semiconductors employed in power conversion circuits by re-examining the descriptions given in Fig. 1.9b.

1.7.2.1 Power losses by power switches in power conversion electronics

Figure 1.11a is redrawn from Fig. 1.9b, and it describes a model circuit topology to feed a diode-clamped inductive load from a DC source by controlling the switching operation of the power switch formed by a power semiconductor. The circuit can be considered as the basic building block of a multiphase inverter systems bridge configuration, explained in later chapters. The associated parasitic

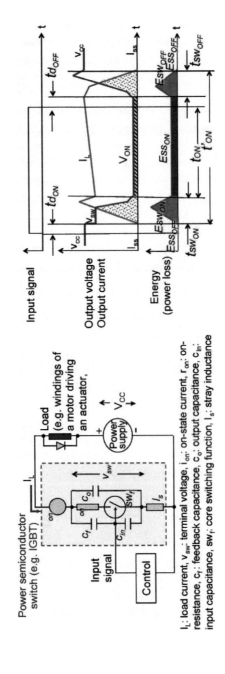

Figure 1.11 Basic switching operation of a power device.

components in the circuit and surrounding the power switch are also shown. The power switch shown in Fig. 1.11a performs on-off operation by the input signal from the control block and feeds the clamped load by circulating current from the power supply.

The "load" shown is a simplified representation of an actual load fed by an inverter system (e.g., connected winding of an AC induction motor or a DC motor). Figure 1.11b shows the voltage and current waveforms observable at the related main terminal of the power switch in response to a control signal fed at its input terminal. Also shown is a waveform of the power switches per switching pulse–based energy (i.e., a representation of power loss) obtained by multiplying instantaneous values of voltage and current at the power switches' main terminals; it can, therefore, be expressed as follows:

$$E = vi \tag{1.8}$$

Considering that the switching operation of Fig. 1.11b is repeated periodically on an on-off cycle having t_{OFF} as the off state time, the total time period can be expressed as follows:

$$T = t'_{ON} + t_{OFF} = t_{swON} + t_{ON} + t_{swOFF} + t_{OFF} \tag{1.9}$$

In this operation, the average power loss $P_{L(ave)}$ generated by the power switch can be given by the following equation:

$$P_{L(ave)} = \frac{t_{ON}}{T} I_{L(ave)} V_{ON(ave)} + \kappa \frac{t_{swON} + t_{swOFF}}{T} V_{CC} I_{L(ave)}$$
$$+ \frac{t_{OFF}}{T} I_{SS} V_{CC} \tag{1.10}$$

where κ is a coefficient derived from the transient locus of the switching operation. The first term in Eq. 1.10 represents the average on-state power loss, the second term represents the average switching-transient average power loss, and the third terms represents the off-state average power loss generated by the switch in its periodic on-off operation.

The coefficient κ in Eq. 1.10 has different values depending on the locus of the switching voltage and current waveforms. In the case of a perfectly inductive load, κ becomes 1/2 as the voltage and current crossover region during switching transient extends to maximum ("hard switching" case). Alternatively, in the case of a perfectly capacitive load switching, κ approximates to zero ("soft switching" or "ideal" switching case). And, for a purely resistive load switching, κ is approximated to be 1/2 ("linear switching" case). However, in most cases of actual inverter or chopper circuits the transient locus of voltage and current tends to take a shape maximizing their crossover region due to the load being very often nearly inductive. Therefore, in the case of such hard switching inductive load applications, the coefficient κ is taken as 1/2 for power loss calculation. To minimize switching losses for improving power conversion efficiency, effort should preferably be made to achieve a soft switching locus by use of commutating or "snubber" circuits.

Chapter 2

Basic Technologies of Major Power Devices

2.1 Power Device Categories

It has been often said that the family tree of power devices pays attention to the similarity of the structures and production technologies. However, it becomes Fig. 2.1 from the point of view of basic operation.

Power devices can be classified roughly into (i) metal-oxide-semiconductor field-effect transistors (MOSFETs), (ii) transistors (bipolar junction transistors [BJTs]), (iii) p-i-n (pin) diodes, and (iv) Schottky barrier diodes (SBDs). The pin diode group represents the power device because it includes insulated gate bipolar transistors (IGBTs), thyristors, gate turn-off thyristors (GTOs), and so on, and occupies overwhelming parts in them. Metal-oxide-semiconductor field-effect transistors (MOSFETs) and SBDs are limited to low-voltage use. The transistor (BJT) is rather a n-i-n (nin) device. So it does not have the ability for high electricity that pin devices have. Therefore, after about 1995, most of the BJTs were replaced by IGBTs.

Power Devices for Efficient Energy Conversion
Gourab Majumdar and Ikunori Takata
Copyright © 2018 Pan Stanford Publishing Pte. Ltd.
ISBN 978-981-4774-18-5 (Hardcover), 978-1-351-26232-3 (eBook)
www.panstanford.com

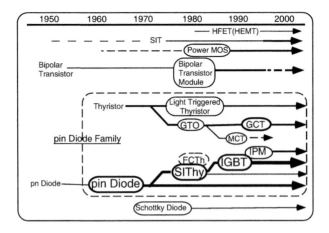

Figure 2.1 Family tree of power devices.

In this family tree, the pin diode, the static induction transistor (SIT), and static induction thyristor (SIThy) are inventions of an honorary professor at the Tohoku University Nishizawa [4].[1] The United States developed the GTO and the IGBT, but it was Toshiba that made a technique worth a product. Fuji Electric (600 V) and Mitsubishi Electric (1000 V) developed the transistor (BJT) for the inverter. Hitachi Ltd. developed the silicon (Si) MOSFET for the high-power device of the cellphone, replacing the high-cost gallium arsenide (GaAs) devices. The gate commutated turn-off thyristor (GCT) was the product that Mitsubishi Electric and ABB codeveloped.

2.2 Key Semiconductor Operation Principles

2.2.1 Essence of the Power Device

The essence of the power device is its switch function, which can go back and forth between on-state and off-state. The ideal device

---

[1] However, Hall and others of the General Electric (GE) Company made the pin structure Ge diode in 1950 as a result [5]. The Si diode was made in 1955 [6]. Nishizawa and others produced SIT and SIThy experimentally in 1974. Ahead of the GTO, the SIThy already had the turn-off ability.

behaves like a metal bar in the on-state and acts as an insulator in the off-state.

As a device near it, a mercury rectifier tube exists represented by a gas discharge. Gas in the discharge tube is usually an insulator, but it turns into a conductive plasma when discharge happens, causing a large current to flow in it.

The free electron's motion breeds an electric current in the semiconductor. This can be increased or removed by adding the forward or reverse voltage. The region's vanishing "free electrons" can become excellent insulators, offering a resistance of more than several hundred MΩcm with silicon, for instance.

As a result, the semiconductor has specific characteristics.[2]

- It can change artificially between a conductor and an insulator.
- A "hole" exists with the free electron.

Because holes and free electrons can exist together in a semiconductor having positive and negative charges, "a plasma state" could be realized, like a gas discharge tube. The inside of the semiconductor can maintain a very high density electric charge by increasing free electrons and holes forming the plasma by an equal amount.

The following two abilities are necessary for a switch:

- Interrupting an electric current
- Maintaining the voltage after turn-off

It becomes hard to use the normal mechanical switch of even around 40 V in a direct current (DC) source. It's because the contact metal melts as an arc current continues when the point of contact is opened. However, this problem does not occur till a value of several

---

[2] These are explained as follows traditionally:

   (i) Electric conductivity is between a metal and an insulator.

  (ii) Electric conductivity increases with temperature.

 (iii) A photoinduced current flows.

However, these were definitions before 1930, when they were only explained phenomenologically. After the transistor appeared, better definitions became essential. In addition, the semiconductor named "Halbleiter" has been in use since 1911.

hundred volts is reached, because an arc disappears at an alternating point in the alternating current (AC) voltage.

The ability to maintain the voltage, as mentioned in the second point above, is usually expressed in a voltage value at which a leak current begins to flow greatly. However, a power device is expected to have abilities to withstand a high voltage, while keeping a large current flowing through it, and also to subsequently turn off safely, ceasing the current flow. The expansion of the limit (safe operating area [SOA]) of this ability—the first point above—has been the biggest problem of power device development. Therefore, the power device was in use through the following three phases:

- Devices without the turn-off process (diode)
- Devices controllable with an AC power line (thyristor)
- Devices that can intercept a DC power line (transistors and GTOs)

And devices in a pin diode series have been used predominantly in devices using more than hundreds of volts. Thyristors, GTOs, and IGBTs belong to this group.[3]

By the way, an IGBT can realize a switch function in itself. However, in the BJT a two-stage Darlington connection is necessary to get the practical on-voltage for 600 V-class and a three-stage Darlington connection is needed for 1200 V-class.[4] In addition, in a GTO, a snubber capacitance C (about 1 μF per 1 kA)[5] and the insertion of the inductance[6] to the main circuit are necessary.

---

[3] From 1980 through 1995, bipolar transistor (BJT) modules, which did not belong to this group, were used widely. In addition, for high-speed operation, the series-parallel connection of MOSFETs is used, too.

[4] Refer to Fig. 1.4, Section 1.3, Chapter 1. The on-voltage increases by about 0.7 V at each stage.

[5] Diverting the main current at the high-voltage-outbreak period in the L-load off-operation, it's able to reduce the GTO's stress. If this capacitance is small, the GTO is easily destroyed. In addition, the electricity saved in capacitance is consumed by the snubber resistance during the GTO's on-operation.

[6] Under a short-circuit condition, its impedance controls the rate of the rise of the main-circuit current to a low value, preventing the device from latching and maintaining gate controllability for a defined period of time pulse operation. For not using this, a fuse is necessary to be inserted, as in the case of a thyristor, to protect the power supply unit.

Although the thyristor can intercept the DC power supply by using a commutation circuit, it becomes a large system because it consists of an auxiliary thyristor and a

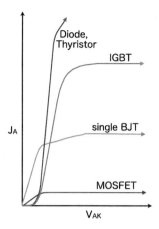

Figure 2.2 Rising characteristics of each device (schematic view).

Therefore, the switching element changes from a thyristor+commutation circuit to a GTO, a BJT, and finally an IGBT. But this priority order does not arise in the following situations:

- Use of a GTO for applications requiring a large current (J_A) that a BJT cannot cover by its rating
- Use of a BJT in the early years of IGBTs when its on-state voltage value was characteristically larger than that of the BJT
- Use of a GTO or a thyristor for applications requiring a large current beyond the coverage possible by IGBTs

Even a GTO could not reveal the potential of a thyristor, and a thyristor-type device reached the completion form as a GCT in 1996.[7]

Concerning BJT, the 600 V and 1200 V products were commercialized as the module structure in 1980 and 1982, respectively. Since then there has been progress of the package, but the semiconductor chip was completed at that point. Along with

---

resonant circuit composed of a inductor and a capacitor, which can handle the main current. Furthermore, extremely delicate control is necessary to avoid commutation failure.

[7] A GCT is a GTO that has a structure that allows a whole cathode current to flow to the gate by lowering the impedance of the gate drive circuit to the extreme. Its semiconductor element is basically the same, and improvements of its package and the drive system allow snubberless operation of a thyristor. (Refer to Section 2.6.2.)

the realization of the control technology by the technological advancement of the central processing unit (CPU), the practical use of the BJT module was indispensable, so an inverter technology to the field of power electronics spread rapidly.[8]

The characteristics of the IGBT became of the same level as those of the BJT, with a 600 V product in about 1993. And after about 1995, the IGBT began to replace the BJT module, with a product of more than 1200 V. The IGBT was put to practical use in a 4.5 kV product afterward and began to replace a part of the GTO. The progress continued in all voltage ranges until about 2005, and it might be considered that the IGBT had almost matured by about 2010.

Outside the field of these power devices, there are unipolar devices that use only the free electrons as charged carriers, such as SBDs, MOSFETs, junction field-effect transistors (JFETs), and heterostructure field-effect transistors (HFETs).

The current control mechanism of the power MOSFET is the same as the one used in an integrated circuit (IC), and the high-voltage operation is the same as that of the bipolar devices described above. Large-current operation is simply achieved by connecting a huge number of small cells in parallel. In addition, a silicon SBD of more than 200–300 V has the fault of a large amount of current leak, especially at a high temperature. Because these devices operate as resistances in an on-state, they are unsuitable for high-voltage use.

It is limited to a Si device due to the fact that a bipolar element is superior in the field of high voltage. In the silicon carbide (SiC) device, a topic of recent years, the SBD and MOSFET are exclusively used.

2.2.2 Characteristics of the Semiconductor

It is difficult to explain concisely the two major characteristics of the semiconductor and the theory of the semiconductor element that was described in the last paragraph. This is because quantum mechanics denying common sense is necessary for understanding the semiconductor.

---

[8]As for the BJT module, substitution by the IGBT advanced rapidly in the late 1990s and the domestic production of the BJT ended in the early 2000s.

Furthermore, the traditional semiconductor theory intends for low-voltage, low-current-density operation or high-frequency operation, which were possible by the semiconductor element when it was announced. There has not been enough consideration of the phenomenon of the power device such as the effect of a long operating area or high-current-density operation under a high electric field.

About the power device, typical of which is a pin diode, there has not been enough consideration of the phenomenon of high-current-density operation under a high electric field in a long operating area.

Anyway, the simple image in which electrons expressed as • and holes expressed as ○ move about in an element is more useful than a profound theory for understanding of the operation of the semiconductor device. Even so, anyone would want to understand the primitive questions of why the "hole," explained as the hole of an electron, can behave as an electron and why holes and electrons can coexist in a huge density.

By the way, the atoms of Si and germanium (Ge) are combined in form, as shown in Fig. 2.3. In the figure, atoms are expressed in the size 1/5 units, but the real atoms are in contact with each other. It seems to be complicated, but the atoms of the neighborhood are connected as in Fig. 2.4. Consider that the chains connected to A-B-C-D in a plane constitute a crystal in Fig. 2.3, combining in length and breadth as shown in Fig. 2.5.

The reason silicon and germanium have such a crystalline structure is that they are tied in "covalent" bonds. And many materials consisting essentially of covalent bonds reveal the inherent

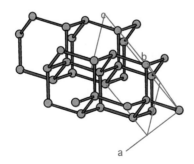

Figure 2.3 Lattice points of a silicon crystal.

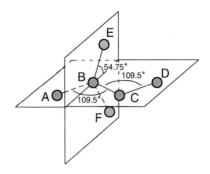

Figure 2.4 1D chains of silicon atoms.

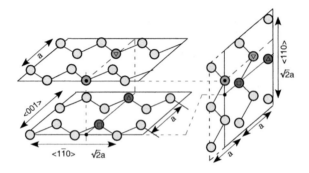

Figure 2.5 Silicon lattice constitution by the 1D linkage group.

characteristics of a semiconductor.[9] If an electron's neighboring orbitals have different spins and their orbital forms are similar,[10] the covalent bond would be easily realized as a kind of orbital resonant, the author thinks.

A covalent bond in a silicon crystal can be expressed as in Fig. 2.6, where the b-electron and the c-electron in each outermost shell goes around each atom B and C, respectively, in the reverse direction.[11]

---

[9] Si, Ge, carbon, graphene (which is one layer of black lead), the nanotube (which is curled-up graphene), some organic molecules, GaAs, SiC, and gallium nitride (GaN) show semiconductor properties. The covalent bond is the strongest chemical bond.

[10] The same kind of atoms or the nearest low number in the periodic table.

[11] Although there is no orbital in quantum mechanics, in this figure the spin ($\pm\hbar/2$) is expediently expressed by the rotating direction. The ▽ indicates the covalent orbital direction, and the ▼ means a covalent orbital and its direction. The pair of • indicates covalent electrons.

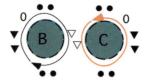

Figure 2.6 Combination hands of a quarter covalent bond.

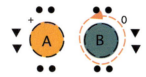

Figure 2.7 Lacking orbital of a silicon atom pair (hole).

In a crystal, all electrons in the outermost shell consist of the spin pairs with the neighboring outermost shell electrons. Because these bonds are equivalent each atom lies in the center of a regular tetrahedron (Figs. 2.3 and 2.4)

In this silicon crystal, it seems reasonable to consider a hole as the Si-atom pair where a covalent bond electron is lacking (Fig. 2.7) and a free electron as the Si-atom pair where an unpaired electron is attached (Fig. 2.8). This pair of atoms can move in a crystal synchronizing with the electron orbital transfer from an atom to its neighbors, accompanying by cut-off of a covalent bond and formation of another covalent bond anew by recombination.[12] These electron orbitals thermally perform a random walk in the crystal, and there is no difference between holes and free electrons. But the center of the hole and the free electron shifts along the electric field and the opposite direction, respectively, when an electric field is applied. This is the reason why holes act very similar to free electrons[13] except the electric charge.

Holes and free electrons have the following characteristics:

- Huge holes and free electrons of almost the same order as that of the semiconductor atom are able to exist.

---

[12] Therefore, mobility decreases in the distracting crystals.
[13] For instance, both of them have the same mass basically.

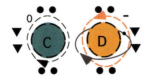

Figure 2.8 Excess orbital of a silicon atom pair (free electron).

- A hole has a higher energy than a free electron. The difference is called an energy gap E_g and corresponds to nearly twice that of the orbital energy.
- The pair generation corresponds to the process where a couple of silicon atoms, as in Fig. 2.6, is transformed to what is shown in Fig. 2.7 and Fig. 2.8. And the recombination is its reverse process.

2.2.3 p-Type and n-Type Semiconductors

The previous section dealt with pure silicon, which is called an "intrinsic" semiconductor. However, semiconductor devices usually use the p-type and n-type semiconductors, where the trivalent atoms, typically boron, and the pentavalent atoms, typically phosphorus, are doped, respectively.

For instance, if a boron atom replaced a silicon atom, Fig. 2.9 would be realized. Therefore when the outermost shell electrons of surrounding Si atoms are transferred to boron atoms, the situation changes to that shown in Fig. 2.10, which is almost the same construction as Fig. 2.6 except for the boron's negative charge. And if a phosphorus atom replaced a silicon atom, Fig. 2.11 would be realized. Besides, if an outermost electron of the phosphorus atoms moved to the surrounding atoms,[14] Fig. 2.12 would appear, which is almost constructionally same as Fig. 2.6 except for the phosphorus's positive-negative charge.

The number of holes generated is almost the same as the number of silicon atoms replaced by boron atoms, and this region is called a p-type semiconductor. Similarly the number of free electrons generated is almost the same as the number of doped phosphorus

---

[14] One of the surrounding Si-atom pairs turns into a free electron (Fig. 2.8).

Figure 2.9 Electron configuration around a boron atom.

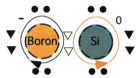

Figure 2.10 Boron atom ion after generating a hole.

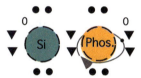

Figure 2.11 Electron configuration around a phosphorus atom.

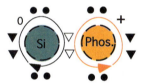

Figure 2.12 Phosphorus atom ion after generating a free electron.

atoms, and such a region is called an n-type semiconductor. Although energy is needed to transfer the outermost electron between a boron or phosphorus atom and its surrounding semiconductor atom, it is easily supplied by the thermal motion for the case of a boron or phosphorus atom in silicon.[15]

These holes and free electrons caused by boron atoms or phosphorus atoms are not distinguishable from those generated in

---

[15] Only such materials whose energy difference for transferring the outermost electron is small could be effective dopants for making p-type or n-type semiconductors.

the intrinsic semiconductor. For these carriers to move around in the semiconductor crystal rapidly, the covalent bond network in it must not be collapsed. So the doping atom ratio is restricted to small. Boron atoms and phosphorus atoms are called "impurities" because they disturb the semiconductor crystal lattice.

In the p-type semiconductor not only holes but also free electrons exist in very small amounts. And in the n-type semiconductor, holes exist in very small amounts. This can be understood by considering that holes and free electrons are generated even in the intrinsic semiconductor by thermal motion.

By the way, the product of a hole density n_h and a free electron density n_e in an arbitrary semiconductor region in the thermal equilibrium keeps constant irrespective of the doping impurity amount (Eq. 2.1). The value C_{he} is a function of the temperature, which is 1.17×10^{20} cm^{-6} at 25°C for the silicon, and it's increased by approximately 240 thousandfold (i.e., 5 orders of magnitude) at 125°C. In the intrinsic semiconductor the n_h and the n_e are equal. So, these values are expressed as n_i and called "intrinsic carrier density."

$$n_h n_e = n_i^2 = C_{he}(T) \tag{2.1}$$

$$n_i : \text{Intrinsic carrier density}$$

$$C_{he}(T) : \text{Function of the temperature}$$

It might seem curious, but a resemble relation between hydrogen ion concentration [H$^+$] and hydroxyl ion concentration [OH$^-$] exists in a water solution. That is Eq. 2.2 is valid independent of the degree of acidity or alkalinity.[16] Both Eqs. 2.1 and 2.2 conform to the same principle and are called the "law of mass action."

$$[\text{H}^+][\text{OH}^-] = 10^{-14} \ (\text{mol}/\ell)^2 \quad (\text{at 298 K})$$

$$= 3.6 \times 10^{27} \ \text{cm}^{-6} \tag{2.2}$$

These laws are generally valid between each density [A], [B], and [AB] (Eq. 2.4) when the reaction where A and B make a compound AB and its reverse reaction are in equilibrium (Eq. 2.3). In the semiconductor A and B correspond to the hole and the free electron and AB is equivalent to the semiconductor atom pair connected by

---

[16]The neutral solution corresponds to pH $= 7$ because it is defined that pH $\equiv -\log[\text{H}^+]$ pH.

covalent bonding. That is, holes and free electrons can be treated as different kinds of ideal gases[17]

$$A \cdot B \Leftrightarrow AB \qquad (2.3)$$

$$[A][B] = C_{AB}(T)[AB] \qquad (2.4)$$

In this way, a hole would be regarded to be of the same rank as a free electron, not as the simple hole of an electron. Then it seems appropriate that the hole and the free electron are expressed as ○ and ● in semiconductor devices.

2.2.4 Potential Barrier between Regions Having Different Impurity Concentrations

Considering the area where a phosphorus-atom-rich n^+-region adjoins a poor concentrating n^--region, free electrons in the n^+-region would move to the n^--region by their thermal motion, as shown in Fig. 2.13. Then an electric field that forced the free electrons to return would appear between them and the leaving phosphorus + ions. If there were no outer bias, the thermal motion and the drift motion of the free electrons would be balanced and reach a thermal equilibrium condition with a certain transient length.

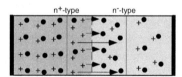

Figure 2.13 Electric lines and impurity ions near the $n^+ - n^-$ boundary (●: free electron; +: phosphorus ion).

Also considering the boundary area where a phosphorus-atom-rich n-region adjoins a boron-atom-rich p-region, free electrons and holes would, respectively, move to the lower-concentration region by their thermal motion, as shown in Fig. 2.14. Then an electric field that forced the free electrons and holes to return would appear between them and the leaving phosphorus + ions and boron − ions.

---

[17] Of course, it's restricted where charge neutrality is established.

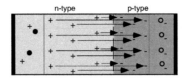

Figure 2.14 Electric lines and impurity ions near the n-p boundary (o: hole; -: boron ion).

If there were no outer bias, the thermal motion and the drift motion of these charged carriers would be balanced and reach a thermal equilibrium condition with a certain transient length. This situation is fundamentally the same as Fig. 2.13.

Now, it's well known for the ideal gasses that the Boltzmann distribution (Eq. 2.5), where N_a and N_b are the densities in the neighboring regions a and b and ΔE_{ab} is the energy difference between them, are generally valid in a thermal equilibrium.

$$\exp\left(\frac{-\Delta E_{ab}}{kT}\right) = \frac{N_a}{N_b} \qquad (2.5)$$

N_a, N_b : Particle densities in the a- and b-regions

Both holes and free electrons are recognized as different kinds of ideal gasses, as mentioned before. So free electrons in the different density regions feel a certain energy difference ΔE and holes feel the $-\Delta E$ because $\Delta E = q\Delta V$. The relation, in which the hole density n_h is inversely proportional to the electron density n_e, is very consistent, as shown in Eq. 2.1.

The material has a particular energy depending on the state and expresses the value in Fermi level E_F in the field of semiconductors.[18] If different objects are in a thermal equilibrium state close to each other, the Fermi level E_F of both varies in only the work function W.[19] For instance, the Fermi level E_F has steps ($E_{F,M} - E_{F,S}$)

---

[18] Sometimes it's called Fermi energy, but we are calling it Fermi level in this book because it might express the electronic kinetic energy $m_0 v_F^2/2$ where it has a Fermi speed v_F. It's called so for the electronic maximum energy that there can be to a certain object in quantum mechanics. It is originally synonymous with the thermodynamic "chemical potential" μ, meaning the energy (depending on the chemical bond and electric potential) of the constitution particle peculiar to the region where it exists.

[19] Energy necessary to take an electron out from a material and to keep it away at the infinite distant point.

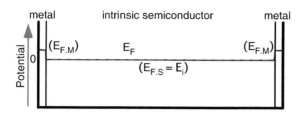

Figure 2.15 Electric potential in a semiconductor with electrodes in thermal equilibrium.

at the boundaries of metals and an intrinsic semiconductor, as shown in Fig. 2.15.[20]

However, in the semiconductors into which ions of the p-type or n-type impurity were introduced, free electrons and holes come under an impurity-ion-related influence different from what they experience being in a pure semiconductor. For instance, when the right side of Fig. 2.13 was an i-region, it's easily understood that the electric potential of the n-region was more than that of the i-region and the electric potential of the p-region was lower than that of the i-region. On the other hand, the Fermi level of the semiconductor does not change because it is an energy level of electrons restricted in the outermost orbital even if p-type or n-type impurities are introduced. Namely, the Fermi levels $E_{F.n}$ and $E_{F.p}$ of the n-region and the p-region do not change from the intrinsic semiconductor (E_i in Fig. 2.15).

These changes in the electric potential are expressed using potential factors that are not effected by introducing the impurities. As these factors, energy levels of the conduction band E_c and the valence band E_v and the middle value of those E_i are available, but E_i is usually used. Then the relations shown in Fig. 2.16 and Fig. 2.17 appear in the situation without the outside voltage (thermal equilibrium state). Because this potential does not appear outside if the electrodes are constructed of the same metal, it is called the "inner potential",[21] particularly in the semiconductor device.

---

[20] E_i refers to the Fermi level E_{FS} of the intrinsic semiconductor conventionally.
[21] The value decreases in order of n$^+$-region > n-region > p-region.

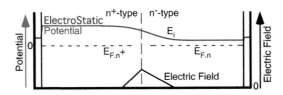

Figure 2.16 Electric field and inner potential near the n⁺-n⁻ boundary (in thermal equilibrium).

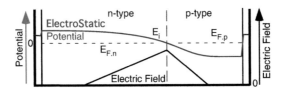

Figure 2.17 Electric field and inner potential near the n-p boundary (in thermal equilibrium).

The inner voltage is estimated from the free electron or hole density difference using Eq. 2.6, which is reduced from Eq. 2.5. q' is equal the elementary electric charge q for the hole and is equal to $-q$ for the free electron and V_{ab} means the a-region's voltage on the basis of the b-region.

$$V_{ab} = \frac{\Delta E_{ab}}{q'} = \frac{-kT}{q'} \ln\left(\frac{n_a}{n_b}\right) \tag{2.6}$$

n_a, n_b : Charged carrier density (n_h or n_e) in a- and b-regions
q' : Electric charge under consideration

For instance, when the impurity concentration of the n⁺-region is 10^5 times larger than the n⁻-region, the former's potential is about 0.3 V higher than the latter's because the free electron densities are almost the same as the n-type impurities. Besides, when impurity densities are expressed as N_A and N_D for the p-region and n-region, respectively, the inner voltage V_{pn} of the n-region on the basis of the p-region is calculated by the Eq. 2.7. It's because the hole density $n_a = N_A$ in the p-region and $n_b = n_i^2/N_D$ in the n-region from Eq. 2.1. If $N_A = 10^{18}$ cm^{-3} and $N_D = 10^{14}$ cm^{-3} then $V_{pn} \approx -0.7$ V. So this p-region's inner voltage is about 0.7 V smaller than that of the n-region.

$$V_{\text{pn}} \approx \frac{-kT}{q} \ln\left(\frac{N_A}{n_i^2/N_D}\right) \qquad (2.7)$$

N_A, N_D : Impurity concentrations of p- and n-regions

2.3 Basic Operation of Power Devices

A power device gets the voltage-holding ability by having a high resistive region as an insulator. And its current-conducting ability can be expressed as Ohm's law (drift current) on the whole. However, the diffusion current becomes dominant under a low-current operation.

2.3.1 Reverse Voltage Blocking

Figure 2.18 shows the typical reverse I–V characteristics of semiconductor devices. "A" and "K" mean an anode and a cathode, respectively.[22] The leakage current and the breakdown voltage are the indicators for the voltage-holding ability. Generally, the former increases exponentially and the latter decreases gradually in a higher temperature. For superior reverse characteristics, the original, namely no doping, semiconductor material must be a good insulator.

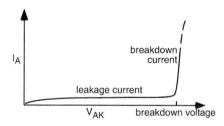

Figure 2.18 Typical reverse I–V characteristic.

The reverse characteristic of power devices is desired not only to prevent current flow but also to withstand the high voltage, keeping a large current flowing. These characteristics

---

[22] The "K" was named after the German "cathode," not to be confused with "coulomb."

might be called "the static withstanding voltage" and "the dynamic withstanding voltage," respectively. The static withstanding voltage of semiconductors can be estimated by a leakage current and an impact ionization coefficient.[23]

The leakage current of high-voltage devices is mostly caused by the generation charges in the voltage-holding region and is expressed as Eq. 2.8.[24]

$$J_R \approx q \frac{n_i}{\tau_g} L \tag{2.8}$$

τ_g : Lifetime

L : Length of the voltage-holding region

The intrinsic carrier density n_i is the hole and free electron density in a thermal equilibrium where holes and free electrons are in a situation that their generation and recombination are balanced in the thermal motion of them and the surrounding semiconductor atoms.[25] τ_g is a time constant of the generation phenomenon and called "lifetime."

To lower the leakage current of high-voltage devices, semiconductor materials whose n_i is small and lifetime τ_g is long must be used. A large-energy-gap E_g[26] material for the smaller n_i and a very pure material for the longer τ_g are needed.

Furthermore, it is desirable for the impact ionization coefficient to be small to get a high breakdown voltage. Pair generation of a hole and a free electron occurs when they collide with a semiconductor atom because of acceleration in a high electric field as well as because of thermal motion. The impact ionization coefficient is defined as the pair generation rate per unit length along the running path.

This ionization action enlarges a leak current source, which is holes or free electrons generated by thermal motion or the irradiation of cosmic rays, while they run to the edge of the device. As the impact ionization coefficient increases exponentially

---

[23] On the other hand, an insulation resistance and a dielectric breakdown field are the measures for the insulator.

[24] At a high temperature, the diffusion leak current ingredient must be considered in addition to this. (Refer to Section 2.5.4)

[25] n_i increases rapidly with temperature and is very small in a wide bandgap (WBG) semiconductor as $n_i \propto \exp[-E_g/(2kT)]$. Also refer to Eq. 2.1, in Section 2.2.3.

[26] E_g means the particular energy difference between the hole and the free electron.

Basic Operation of Power Devices | 45

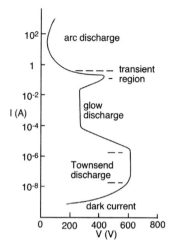

Figure 2.19 *I–V* characteristics of the gas discharge tube.

with the electric field, the seed charge of a leakage current could generate many more charges passing through the device in a high voltage. Then the charged carriers multiply like rabbits above a certain voltage. Moreover, because the impact ionization is also proportional to the charged carrier density, the leak currents increase rapidly at a certain voltage recursively.[27] That is the "breakdown phenomenon".[28]

When the breakdown phenomenon occurs, densities of holes and free electrons increase in the device and the reverse current could continue even if seed charged carrier are cut off when their densities are larger than the seed's densities at the beginning of the breakdown phenomenon.[29] This continuation current spreads in the whole cross section of the n^--region and is stable in spite of the radical phenomenon with the large density of charged carriers, and its density J_R is determined uniquely by the applied voltage V. The semiconductor device is not destroyed at this stage unless its temperature rises enormously.

---

[27] This positive feedback phenomenon is likened to a snow slide—an avalanche—and is called an "avalanche phenomenon."

[28] It corresponds to the "Townsend discharge" of the gas discharge tube in Fig. 2.19.

[29] In the gas discharge tube, the secondary electrons, which are generated by a gas atom ion colliding with the cathode, become the new electron source.

This situation corresponds to the "glow discharge" of the gas discharge tube in Fig. 2.19.[30] When currents increase in the discharge tube, an unstable "arc discharge"[31] happens through the stable glow discharge. In the semiconductor device a similar phenomenon occurs if the breakdown current increases.[32]

Germanium (Ge) is an inferior material compared to silicon (Si) for use in high-voltage power device fabrication. On the other hand, silicon carbide (SiC) is a considered a superior material for said device application compared to Ge and Si. This tendency corresponds with their energy gap E_g. It is because the values of the intrinsic carrier density n_i and the coefficient of the impact ionization get smaller according to an increase of the energy gap E_g of the material.

Figure 2.20 shows a typical example of how to maintain a high voltage with a semiconductor device. A diode like this, which has the long low-impurity region between the p-region and n-region, is called a pin diode. The "i" in pin means "intrinsic," which means "does not include impurities," but it is conventionally used when impurities are relatively low in concentration, namely, when resistivity is fairly high. In the real power device, where the n^--region is more than several dozen Ωcm, it is often called the i-region.

If the i-region is an insulator with a length L, a uniform electric field of V_R/L takes place when a reverse voltage V_R is added to a pin diode. By this electric field, the free electron of the n-region and the

---

[30]The "dark current" in Fig. 2.19 consists of charged carriers that were generated by the incident charged particles, such as cosmic rays, and correspond to the leak current of semiconductor devices. The Townsend discharge is the situation in which it causes avalanche multiplication by a part of the cross section of a discharge tube. The glow discharge is considered to occur because the Townsend discharge spreads through the whole cross section equally and the current flows stably. The mercury rectifier uses this glow discharge at on-operation.

[31]Not only electrons but also metal ions, which evaporate from an anode, take a current.

[32]Moreover, the discharge properties change significantly depending on the length and gas pressure of the discharge tube. For instance, when the gas pressure becomes lower, the Townsend discharge begins in several fA and the current range of the glow discharge shortens. Since there are many similarities between gas discharge tubes and semiconductor devices, the $I-V$ characteristics of both resemble closely.

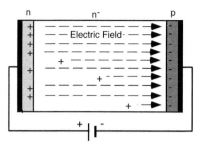

Figure 2.20 Voltage-holding situation of a pin diode.

hole of the p-region are kept away from the i-region and positive ions and negative ions of impurities are left on each part of the n- and p-regions that are in contact with the i-region. This state is basically the same as the situation in which voltage is applied to a capacitor.

In Fig. 2.20 and Fig. 2.21, the i-region is not an insulator but the n^--region is one, actually. In the n^--region at this time, a free electron is removed by an electric field and in this situation positive impurities ions are distributed sparsely. Unlike the case of the insulator, some electric lines of force begin to start in the middle of the n^--region, and not the n-region, of the cathodal side. Because the electric field strength (EF) is proportional to the density of the electric line of force, its distribution shows a gentle slope, as in Fig. 2.21.

In addition, with the semiconductor device using only one of the holes or free electrons, such as the MOSFET, electric lines of force that start from the n^--region increase to get a priority to current-

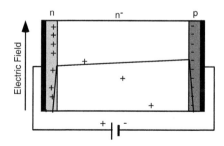

Figure 2.21 Electric field distribution of a pin diode.

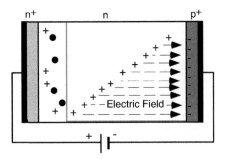

Figure 2.22 Voltage-holding situation of a p$^+$nn$^+$ diode.

carrying capacity, and as for the distribution of the EF, it is with a steep grade. An electric line of force beginning in the n$^+$-region of the cathodal side would disappear, as shown in Fig. 2.22 and Fig. 2.23. At that time, the maximum EF for the same holding voltage inevitably increases to more than that of pin diodes, as shown in Fig. 2.21.

Many documents explain that a breakdown phenomenon, as drawn in Fig. 2.18, occurs when an EF reaches a specific critical value EF_C. It almost applies to the situation of Fig. 2.23, on which the degree of leaning of the EF distribution has a big impact. This is because there is strong EF dependence for collision ionization action. However, it is difficult to let the specific critical value EF_C support in the situation of Fig. 2.21, where the leaning of EF is small and must consider an integrated value of the collision ionization coefficient across the whole n$^-$-region. Therefore, the resistivity of

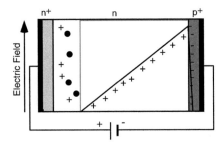

Figure 2.23 Electric field distribution of a p$^+$nn$^+$ diode.

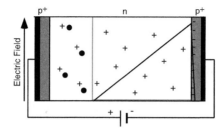

Figure 2.24 Basic structure of a backward withstanding voltage device.

the n⁻-region becomes large and the maximum of EF, which brings about the breakdown phenomenon, tends to become small.[33]

It's necessary primarily to lengthen the n⁻-region to raise the static withstanding voltage because the holding voltage is equal to the area of the EF distribution. And it is desirable to lower the impurities density of the n⁻-region. The real silicon devices of more than several hundred volts are nearly as thick as the n⁻-region, with a 10 μm per 100 V withstanding voltage, and a situation as shown in Fig. 2.21 is realized.

By the way, with the device structure of Fig. 2.20 and Fig. 2.22, it's impossible to withstand the reverse voltage when a power supply is connected in the opposite polarity. The structure with the p^+np^+ three layers that uses the p⁺-region instead of the n⁺-region in Fig. 2.22 is thought to keep withstanding voltages in the bidirectional polarity of the power supply (Fig. 2.24). But it is necessary to thicken the n-region considerably than n^+np^+ diode of Fig. 2.22 because leak currents increase if the hem of the electric field existing area approaches the p⁺-region, at a high temperature in particular. Although this structure cannot flow the forward current, it's the basic structure to get nearly the same withstanding voltage bidirectionally.

In this way, the static withstanding voltage is decided on the basis of the thickness and resistivity, or impurities density, of the n⁻-region in conformity with a simple principle. However, concerning the peripheral portion of device chips, this principle is not easily

---

[33] It can become $EF_{max} \leq 1$ MV/cm with the ultra-large-scale integrated circuit (ULSI) but is about $EF_{max} \approx 0.2$ MV/cm with devices of several thousand volts.

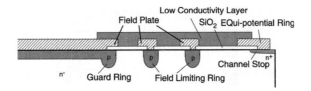

Figure 2.25 Typical peripheral structure of a planar chip.

adoptable. Long-term reliability becomes, in particular, difficult with a high-withstanding-voltage device. Basically, it is necessary to avoid concentration of the electric field, and it has only been validated to etch diagonally the edge, to increase the edge face length, and to cover it with resin. This method is still used to put a large-area thyristor and diode wafer in a large ceramic package (cf. Figs. 2.55 and 2.56).

However, the current practice is to leave the original surface of the planar structure of the silicon covered with its oxidized film. This method, which induces a good production yield and excellent long-term reliability, has become indispensable for the module that mounts many chips in a poor sealing package.

Figure 2.25 shows the typical peripheral structure of the planar device [7]. The important points are (i) to avoid the concentration of the electric field and (ii) to prevent the surface of the n⁻-region surface from turning over in the p-region. The guard ring and field limiting ring are measures of (i) and the equi-potential ring, the field plate on the FLR, and the low conductive layer are provided for (ii). For high-voltage devices, a polyimide resin is sometimes coated on this structure.

Although the voltage-holding ability of the semiconductor device seems to rise endlessly by increasing the resistivity and length of the n⁻-region, heat generation by the leakage current really limits it through the thermal runaway, which increases the temperature recursively. The leakage current gets larger with the voltage rating and the junction temperature T_j. Therefore, today's commercially available thyristors and diodes of 12 kV at $T_j = 125°C$ would be the upper-limit voltage in silicon.

In addition to the static withstanding voltage mentioned before, where no current substantially flows, a voltage-holding ability in the

Basic Operation of Power Devices | 51

large current flowing situation, namely "safe operating area," must be considered. It's treated in Section 2.3.3.

2.3.2 Forward Conducting

The electric current in the semiconductor mainly flows by a "diffusion" and "drift" mechanism.[34] Although many books explain only diffusion current, the drift current in the long n^--region is much more important in power devices. The current density J by the drift mechanism of the hole and the free electron is expressed in Eq. 2.9, where for each drift speed v_h, v_e is referred to in the product of the drift mobility μ_h, μ_e, and an EF (Eqs. 2.10 and 2.11).

$$J = qv_h n_h - qv_e n_e \qquad (2.9)$$

$$v_h = \mu_h EF \qquad (2.10)$$

$$v_e = \mu_e EF \qquad (2.11)$$

$\quad n_h, n_e$: Densities of holes and free electrons

$\quad v_h, v_e$: Velocities of holes and free electrons

$\quad \mu_h, \mu_e$: Drift mobilities of holes and free electrons

$\quad$ EF : Electric field

The Eq. 2.12, which expresses the drift current I is nothing but Ohm's law. Conductivity $(1/R)$ of the conducting area is proportional to the drift mobility μ and charge density n (Eq. 2.13).

$$I = q(\mu EF)nA = q\mu n\frac{V}{L}A = \frac{q\mu nA}{L}V = \frac{V}{R} \qquad (2.12)$$

$$\frac{1}{R} \equiv \frac{q\mu nA}{L} \qquad (2.13)$$

$\quad A$: Cross section

$\quad L$: Length

$\quad I$: Current

$\quad V$: Voltage

$\quad R$: Resistance

To increase the conducting performance of the power device, it's very effective to enlarge a cross section A and to shorten the

---

[34] In addition, there are other kinds of mechanisms, such as recombination, pair generation for leak current, and multiplication of it for breakdown current.

length L of the device as a matter of course. Mobility μ has a value particular to each semiconductor and decreases after each wafer process, mainly by a heat treatment. Therefore, it becomes the key to increasing the hole's or free electron's density n_h and n_e, respectively. For that purpose, four kinds of methods are thought about:

- Increasing impurities (boron or phosphorus) that are introduced into the conducting path.
- Producing an electric field from the outside at right angle to the conducting path and accumulating electric charges at the surface of it.
- Producing an electric field at right angle to the conducting path by a distortion stress in the semiconductor and accumulating electric charges at the surface of it.
- Constructing a structure in which holes and free electrons flow into the conducting area from each side of the semiconductor region neighboring it:
  - This is the most influential method for getting a good performance of the unipolar semiconductor device,[35] which uses only one of the holes or free electrons. However, an appropriate balance is necessary because higher impurities induce lower withstanding voltages, as mentioned in the previous section.
  - This is the mechanism used in the current channel part of a MOSFET. It's for this reason that the on-resistance falls, so as to increase the gate voltage.
  - This is the mechanism to form the current channel layer of a HFET.

A typical example of the fourth point is the pin (actually, pn$^-$n) diode, as shown in Fig. 2.26.[36] In the forward operation, free electrons and holes flow into the n$^-$-region from the n-region and the p-region, respectively. The densities of the holes and free electrons increase equally to $n_i \exp(q V_0/2kT)$ in the whole

---

[35] For instance, JFETs, vertical MOSFETs, and SBDs.
[36] The length of the n-p-region should be short (less than a few μm) theoretically because its performance is usually improved. (Refer to Section 2.4.3)

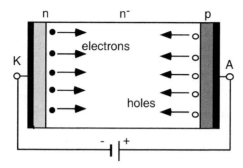

Figure 2.26 Forward operation of a pin diode. (The shorter the n-p-region length, the better the performance.)

n⁻-region, except a large current operation, as explained in Section 2.4.3, where V_O is the voltage across the terminals. Besides, in the small current operation, the potential difference V_d in the n⁻-region could be ignored and the current density J becomes the exponential function of V_O, as in Eq. 2.14.

$$J \propto \exp\left(\frac{qV_O}{kT}\right) \qquad (2.14)$$

And in current operation higher than the rated current of the high-voltage diode, the Ohmic drop V_d in its n⁻-region gets dominant on the external voltage V_O across the terminals ($V_O \approx V_d$). The current density J can be expressed as a product of a charged carrier density n_h, n_e, its drift mobility μ in the whole p-n-region and n⁻-region, and the EF $\approx V_d/L$ of the whole diode, as in Eq. 2.15.[37]

$$J \approx q(n'_h + n'_e)\mu'\frac{V_O}{L} \qquad (2.15)$$

n'_h, n'_e, μ' : n_h, n_e μ in the n⁻-region

L : Length of the whole diode

---

[37] In the electrode adjacency area the operating current density J can be expressed as Eq. 2.16. Moreover, both EF_{p0} and EF_{n0} get closer to V_O/L in a huge current density.

$$J \approx q(N_A \mu_{h0} E F_{p0} + N_D \mu_{e0} E F_{n0}) \qquad (2.16)$$

N_A, N_D : p, n-region impurity concentrations

μ_{h0}, μ_{e0} : μ_h, μ_e of p, n-regions near each electrode

EF_{p0}, EF_{n0} : Electric field of p, n-regions near each electrode

54 | *Basic Technologies of Major Power Devices*

Because it is necessary to lengthen the n^--region and increase its resistivity to get a high withstanding voltage, as mentioned in Section 2.3.1, the current-conducting ability decreases necessarily. On this account, for a silicon device particularly of more than several hundred volts, a structure that has relatively highly doped p-type and n-type regions on both sides, as in a pin diode, to let holes and free electrons flow into the middle of the n^--region from these region is only practical. All the main high-power semiconductor devices, such as thyristors, GTOs, GCTs, SIThys, and IGBTs, possess this structure.

2.3.3 Voltage-Holding Ability with Large Current: SOA

BJTs were not available above several hundred volts until the late 1970s because they were easily destroyed on increasing the operating voltage. They were destroyed without the omen in a step stress test and lost the voltage-holding ability instantaneously. These sudden phenomena in operating stress and time were mysterious because it's hard to regard a temperature rise as the cause, and these have been called the "second breakdown." To use such transistors (BJTs) delicate know-hows in snubber circuits and so on were indispensable.[38]

The prosperity of today's power electronics began with the mass production of the general-purpose inverter in the early 1980s. It is because large electricity transistors (BJTs) that solved the second breakdown of the 600–1200 V class were in use. And, it may be said that historically, along with the process of expanding the practical use of power devices (BJTs, IGBTs, and GCTs) in switching applications, this SOA (an area defined by operable voltage and current where the second breakdown phenomenon does not occur) have also expanded.

> In comparison with the second breakdown, the words "primary breakdown" is not used; merely "breakdown" is used. It means "destruction," but it's a stable phenomenon and the reverse current merely flows continuously. However, it would be named "breakdown" because devices were destroyed while measuring the withstanding voltage at the initial stage of the semiconductor study.

---

[38] Refer to Section 1.3, Chapter 1.

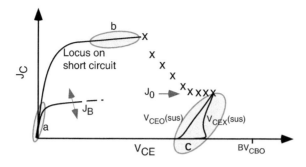

Figure 2.27 Safe operating area (SOA) of a transistor (BJT).

Concerning the second breakdown of transistors (BJTs), a current density J_C and the main voltage V_{CE} were the main factors, and the operating power density ($J_C V_{CE}$) was particularly important as a limit parameter. Therefore, the SOA of the transistor (BJT) can be expressed as Fig. 2.27, where the vertical axis indicates the current density J_C and the horizontal axis indicates the collector-emitter voltage V_{CE}. In this figure, each × expresses a break point. In these breaking points, the relatively low current parts are the results in the L-load turn-off tests and the higher current parts are the results in the load short-circuit tests, varying the base current J_B. This SOA can be regarded as the limit of the voltage-holding ability in the current flowing situation.

A transistor (BJT) holds a stable operating mechanism in each area, a, b, and c, in Fig. 2.27. The area a indicates the operation in a saturation region, where enough base electric current supplied, the area b indicates a load short-circuit operation, and the area c indicates a sustain operation.

> A sustain operation is a situation in which a severe current flows while maintaining the net limit of its withstanding voltage, which is inherent for the semiconductor device itself. In the case of a BJT, it changes depending on the base voltage reading V_{BE}. A state of the base opening is $V_{CEO(sus)}$, and the reverse bias state ($V_{BE} < 0$) is $V_{CEX(sus)}$. When $-V_{BE}$ increases, the low-current part of $V_{CEX(sus)}$ moves to the high-voltage side but the voltage of the maximum current point J_0 does not change.

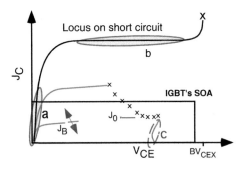

Figure 2.28 Comparison of the SOA between an IGBT and a BJT.

This maximum current J_0 is the value where the free electron density in the n⁻ collector region reaches the impurities density N_D[39] in it in the case of npn BJTs. For BJTs of several hundred volts, this J_0 is dozens of A/cm^2. The BJT is not originally destroyed by an operating current density less than J_0.[40] In addition, the rated voltage of BJT expresses it in $V_{CEX(sus)}$, but the actual breakdown voltage BV_{CBO} between the collector and the base of a normal high-voltage BJT would be approximately 1.3 times.

The SOAs of IGBTs up to about 1200 V are so large that they are not declared clearly and are generally accepted to be expanded over double the rated current at the rated voltage of the device (square SOA covering 2× rated current). Figure 2.28 compares the SOA of an IGBT with that of a BJT having a n⁻-region of the same thickness schematically. The SOA of the IGBT is much larger than that of the BJT. However, the SOA becomes smaller with an increasing rated voltage; the double square SOA is difficult in the 4.5 kV class IGBT.

The IGBT has also a stable mechanism in the saturated operation area a and the short-circuit operation area b of Fig. 2.28. This is because the operating current of the IGBT is limited by the conducting ability of the built-in MOSFET. However, the IGBT cannot exhibit a stable operating area corresponding to the sustaining operation of area c. In an IGBT, the holes account for dozens of percent in a current component in addition to free electrons and

---

[39] $J \equiv q N_D v_s$. v_s: Saturation velocity (≈ 100 km/s).
[40] It is the case of a short time operation, typically less than 10 μs for BJTs, due to which the temperature rise being negligible is not of any concern. Refer to Fig. 2.71.

the density difference $(n_e - n_h)$ has a bigger influence than the density itself. Then, although the operating limit of the npn BJT is easily explained just by considering the free electrons' density n_e and current J_e, IGBT's cannot be described simply.[41]

In addition, the IGBT's rated voltage has only some measurement margin to the net breakdown voltage BV_{CEX} between the collector and the emitter usually and does not have a direct relation with its SOA.

The GTO, which handles large currents, typically more than several kiloamperes, has to withstand a high voltage that is generated by the stray inductance existing in the main circuit where the main current flows at every turn-off operation of the device. Destruction of this stress is the second breakdown of the GTO. However, it is necessary to connect a snubber capacitor in parallel because the GTO cannot endure practical use by itself.[42] GTO's stress could be largely reduced if most of the main current was passed to this capacitor $(\approx\mu F/kA)$ when the anode voltage V_{AK} of the GTO begins to increase.[43] The second breakdown resistance of a GTO greatly depends on the size of the snubber capacitor, and the condition of the voltage and the current where destruction happens in a turn-off process are not determined. Therefore, the figure "safe operating area" is not used in a GTO.

On the other hand, a GCT is a device that improved the GTO, so a turn-off is enabled without the snubber capacitor. Currently, only the GCT specifies the limit value (I_A, V_{AK}) of the second breakdown as the current intercepting capacity in device rating. However, a much huger current flows in a GCT/GTO than in an IGBT in the load short-circuit operation because a GCT and a GTO have no mechanism to confine flowing current, like that built into the MOSFET of an IGBT. Because its short-circuit current largely surpasses the intercepting rating, a GCT/GTO cannot turn off, which leads to destruction

---

[41] Refer to Section 2.9.4.

[42] Refer to the GTO construction in Fig. 1.4, in Chapter 1.

[43] Refer to Figs. 2.52 and 2.53, in Section 2.6.2. The current that flows through the snubber capacitor makes the snubber resistor hot and causes a power loss of several percent. A snubber with the same purpose was necessary for BJTs before about 1980, as mentioned in Section 1.3, Chapter 1.

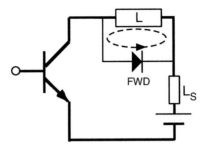

Figure 2.29 Circulating circuit with a freewheeling diode (closed circuit part with L and FWD).

inevitably. In other words, a GCT/GTO does not have short-circuit-withstanding capacity.

Each second breakdown phenomenon of the BJT, IGBT, and GCT/GTO mentioned here contains the turn-off operation of an inductance load L. Besides, because any semiconductor device is not able to handle severe current flowing through an inductive load at switching by itself, a circulating circuit, Fig. 2.29, is usually set up and prevents it from the counter electromotive force by an L-component of the load.[44] Therefore, the inductance L becoming the problem is really a stray inductance L_S of the line except the circulating circuit from the main circuit.[45]

However, to produce a high voltage using an inductance load positively, a circulating circuit or a clamping circuit could not be adapted. And the switching device must turn off the main current by itself while holding the high voltage that was generated by an inductance load. Then the semiconductor device is forced to perform the sustain operation mentioned above. The destruction resistance of this situation is called the "avalanche withstand capacity" without distinguishing the second breakdown from the destruction by the temperature rise in particular. BJTs, IGBTs, and vertical MOSFETs are used for such applications.

For instance, in a BJT, the electrical limit of this operation is the current density J_0 of the upper limit point of the sustain area (c)

---

[44] If there was no circulating circuit, a voltage clamping circuit with a slightly low withstanding voltage might be connected in parallel to the switching device.

[45] The closed circuit is indicated by a dashed line in Fig. 2.29.

in Fig. 2.27. In the operation under this current density J_0 only a temperature rise causes destruction. On the other hand, an IGBT can handle a more severe current than a BJT, but it is necessary to test this in an actual usage condition because the upper limit voltage and current are not expected generally.

A vertical MOSFET has the same breaking point J_0 of Fig. 2.27 as a BJT. However, its value is usually far superior than that of BJT because J_0 is equal to $q N_D v_s$ and its impurities density N_D in the n$^-$-region is usually an order of magnitude greater than that of a BJT. So a temperature rise or an imbalanced operation in the chip becomes a more important factor. Although a MOSFET is a stable device as its conducting area is basically a resistor, a thermal runaway might be possible when it is inserted in a constant current circuit because the resistance value increases monotonously with temperature.

On the other hand, the pin or pn diode has two kinds of second breakdown. The first is an upper limit point of the sustain operation in Fig. 2.27. This breaking current J_0 is the same as that of the BJT, but the sustaining voltage becomes higher than BJT's because there is no current magnifying action that is symbolized as h_{FE}. In other words, the sustain waveform of a diode rises up at the BV_{CBO} point in Fig. 2.27 if the n$^-$-region is the same as the BJT's.[46]

The second of the second breakdown is destruction at the time of the recovery operation.[47] The other second breakdowns, except this operation, are involved in an avalanche multiplication phenomenon of ionizing collision and would be destroyed at the maximum applied voltage V_{AK} when increasing the test condition gradually.

However, it is not unusual for the recovery destruction to occur before the maximum voltage point and at a much lower voltage V_{AK} than the value of the previous test. This is the most desirable second breakdown phenomenon to be analyzed in the present.

The second breakdown of the JFET and HFET (or HEMT) is not analyzed well. Because all conducting areas of these devices are resistances, it was thought to be a basically stable device. However,

---

[46]The avalanche destruction of the well-designed MOSFET is nothing but the destruction of this mode of the pn diode that is built in. In the MOSFET that is not so, $V_{CEX(sus)}$ operation starts at the voltage lower than a BV_{CBO} level. Anyway, all MOSFETs are finally destroyed thermally.

[47]Refer to Section 2.4.2.

it is doubtful whether there is a stable operating mechanism in the situation that charged carriers cause an avalanche multiplication in a high electric field.

Generally, power devices are constructed from a lot of small basic elements that are connected in parallel. And the destruction of the device occurs at the element of the worst operating condition exceeding the element's ability in it. The uniform operation is really necessary to enlarge the SOA and not only to improve the element's performance. The weight of this imbalance problem in parallel operation seems to be much more serious in a high-performance device such as an IGBT.

2.4 Diode Rectifiers

2.4.1 Diode Structures

Originally "diode" was the name of the vacuum tube that has the anode and cathode electrodes but is now used as the name of the semiconductor device of two terminals doing detection action, the same way as the diode tube. By the way, the term "rectifier" is used for a "current flow straightener," for instance, a mercury rectifier or a selenium rectifier.

The history of the diode dates back to a metal-semiconductor rectifier of the 1920s. The selenium rectifier, which was the first power rectification device, was constructed from a cadmium alloy electrode on the polycrystalline selenium fusion-bonded on a metal plate.[48] Although they were 3 orders of magnitude greater than silicon rectifiers, selenium rectifiers had been made from the 1930s until the first half of the 1980s because they were hard to be destroyed in a voltage surge and serial/parallel connections were easy.

The metal-semiconductor rectifier was completed during World War II as a point contact diode of silicon or germanium for the

---

[48]Selenium is essentially a p-type semiconductor. The selenium rectifier was put to practical use from about 1920 and has the ability of nearly 50 mA/cm^2 30 V. Around the same time, the cuprous oxide (Cu_2O) rectifier had begun to be used for small and medium current. These were called metal-semiconductor rectifiers.

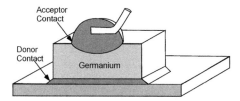

Figure 2.30 Diffused junction diode [8].

detecting device of the radar microwave.[49] However, it was in 1974 that Siemens AG realized the 20 V power device using silicon for microwave use and named it "Schottky diode" after W. Schottky, its researcher.

In 1948 GE sold the small signal diode that had lead welded to Sb-doped n-type germanium in a cylindrical package.[50] In 1950, GE used the plastic package as the small resistor and produced it in large quantities (G-series) [8]. Next, an alloy-treated device in which pellets of indium (In) and Sb were alloyed on both sides of a pure germanium piece was also manufactured (Fig. 2.30).[51]

The power pn diode using germanium had an advantage, a low on-voltage, but its usage was limited to under 80°C and to 200–300 V at the most.[52] Therefore, a silicon diode was considered to be the favorite from the beginning as the power rectifier. When the silicon pn diode appeared in about 1955, at first it replaced the selenium rectifier and came to cover the several kA class mercury rectifiers continuously because it can hold high voltages of even more than 1 kV at 150°C

Silicon pn diodes are used now. As for the high-voltage diode whose rating is more than several hundreds of volts, the pin structure that contains the n^--region of high resistivity between

---

[49] Polycrystalline silicon or an antimony (Sb)-doped pure germanium crystal was used. They were sealed with a cylindrical ceramic and metal electrodes at both ends.

[50] Although a needle of the point contact diode also melted slightly, its needle was basically pushed mechanically.

[51] It became the p^+ i n^+ diode whose p- and n-regions were thick and were highly doped by the p- and n-type impurities, respectively. Even a product was realized having ratings of 185 V withstanding voltage and a 0.5 A consecutive current at 55°C.

[52] In 1952, GE announced the trial manufacture of the diffused junction diode of 150 V 0.35 A (at 55°C) but was not able to realize it.

the p-region and n-region is overwhelming. The reason why it is called the pin diode is that it was theoretically possible for the n^--region to be the genuine, or intrinsic, region that does not include impurities. The concept of a pin diode was raised in 1950 [4][53] and was realized as a silicon diode of several hundred V in about 1960.[54] Furthermore, after 1980, on the basis of various inventions, it was improved to get low-loss properties at high-speed switching for usage as a freewheeling diode (FWD) of an inverter or a rectification diode of a switching power supply.

And a SBD was used frequently next because its forward voltage is about half and its switching speed is much faster compared to that of the pin or pn diodes. This SBD was the modernized metal-semiconductor rectifier that had been used since 1920, and it was finally manufactured for power devices by using silicon as a semiconductor in about 1970. The SBD has a simple structure in which a metal is in contact with a semiconductor surface, as shown in Fig. 2.31. As the contact treatment between the semiconductor and the metal was difficult, it took a lot of time for SBDs to come into practical use. Products of more than 200 V are difficult even now when they use silicon.[55] The SBD of Fig. 2.31 uses n-type silicon, but reverse polarity is provided if p-type silicon is used. However, this p-type SBD is not sold on the market because it shows a slower switching speed and a higher on-voltage than the n-type SBD as a hole's drift velocity is slower than an electron's.

In addition, the high-speed diode that has a Schottky barrier contact at the anode part of a pin diode, as shown in Fig. 2.32, had begun to be used for the 600 V class IGBT module from about 1990 [9]. This structure was called the merged pin Schottky (MPS) diode and was expected to prevent oscillation at the time of low current switching.[56] In Fig. 2.32, to prevent the increase of the

---

[53]The junction diffused diode of GE, which was released in 1950, also had the pin structure unintentionally.

[54]When an epitaxial transistor was made in 1960, it had appeared between the base-collector portion unintentionally. Refer to Section 2.7.1.

[55]It is because the leak current increases and the withstanding voltage reduces, in particular, at a high temperature.

[56]In about 2010, an improved 600 V product (Qspeed diode[TM]), in which trenches are dug, forming the p-region at its chip, was manufactured in the United States. The one using SiC is called a junction barrier Schottky (JBS) diode.

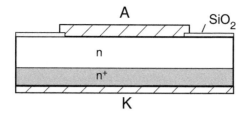

Figure 2.31 Basic structure of a Schottky barrier diode (SBD).

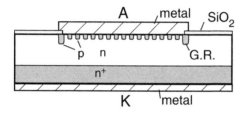

Figure 2.32 Basic structure of a MPS/JBS diode.

leakage current around the Schottky electrode, the guard ring (G.R.) is attached.[57]

Now, the first technical problem of the diode is reducing loss in high-frequency switching because it has pushed forward high-frequency switching in both the inverter and the switching power supply. The next problem is lessening the electromagnetic interference (EMI) noise at the switching operation. It is pointed out that the EMI noise from the diode is generally larger than that from the IGBT or the MOSFET.

About the low loss in high-frequency switching, the reduction of the switching loss is given priority over lessening the on-voltage, partly because the on-voltage of the high-voltage diode has a small ratio for the power supply voltage. When a switching device such as an IGBT is turned off in the inverter circuit or the chopper circuit, the switching loss is caused by the switching device itself. However, the loss when an IGBT is turned on depends on the recovery

---

[57] Almost all high-voltage semiconductor devices with a planar structure adopt appropriately the chip terminal structure that is explained in Fig. 2.25 in Section 2.3.1.

property of the FWD as the IGBT is forced to the load short-circuit operation until the FWD finishes its recovery process. Therefore, the improvement, mainly shortening of the recovery time, of the recovery property of the FWD is strongly demanded.

An SBD shows excellent recovery properties with a low on-voltage for the following reasons:

- The voltage rise in the SBD is about half that in the pin diode.
- Because only either a free electron or a hole is an exercising element like a resistor, electric charges are not accumulated in the operating region, unlike the pin diode.

However, the leakage current of the SBD is orders of magnitude larger than that of the pin diode. In addition, the on-voltage increases necessarily if the current density becomes large in the unipolar operation of the second item, and these tendencies become remarkable if the temperature rises. Furthermore, the high-frequency properties get worse when the resistivity is increased to accomplish a high voltage because holes are generated at the high-resistive n-type semiconductor adjacent to the metal and the SBD reduces its unipolar property. Therefore, although a silicon SBD is used for high-frequency switching lower than about 100 V exclusively, products more than 200 V are difficult.

In case of more than 200 V, where a silicon SBD is not usable, a pin diode was used until about 2000. The pin diode is excellent as a high-voltage diode, but in most peoples' opinion, there is no means to improve considerably the trade-off between a high-speed performance and an on-voltage.[58] If using silicon, the practical voltage of the SBD seems to be restricted to below about 200 V, but its operating area is enlarged to several thousands of volts if a wide-bandgap (WBG) semiconductor such as SiC is used.

Concerning the SBD of SiC, a 600 V product was manufactured in 2000 by Infenion,[59] and 600–1200 V is commercially available now. Because it is expensive, the use range was limited to some

---

[58]However, an improved MPS diode, a Qspeed diode™, of 300–600 V began to be used from about 2010.

[59]It's the split company of the semiconductor section of Siemens.

switching power supply, but the usage area spread to the FWDs of the IGBT module. If a cheap and stable production technology of SiC substrate was established, it would spread widely. A product of several kilovolts has begun to be manufactured now.

But the on-voltage of a SiC SBD becomes the same level as the silicon pin diode because the rising voltage increases along with the E_g. Therefore, it is expected in an FWD where its good recovery properties are utilized or in a high-frequency switching field.

On the other hand, as even the on-voltage of the silicon SBD is not low enough to use for the switching power supply whose output voltage is several volts, a synchronous rectification method in which the MOSFET is synchronized with the switching frequency is widely used. Although it reduces the on-voltage up to nearly half of SBD's (less than 0.2 V), it is necessary to be careful about the negative influence of the parasitic pn diode that is formed inevitably between a drain and a source, as explained in Section 2.8.

2.4.2 Transient Operation of a pin Diode

For the diode used in a switching circuit, not only the off-/on-state but also recovery characteristics just at the on-/off-timing of the pulse operation is very important. Although these characteristics should be called the forward recovery and the reverse recovery, respectively, those are merely mentioned as the transient on-voltage and the recovery usually.

Figure 2.33 shows the forward recovery property. During the transient period after a forward voltage is suddenly applied, there is a delay time t_{fr} before a current begins to flow and shows a forward direction voltage V_{FP} that is higher than the stable on-voltage. This transient on-voltage usually does not cause a problem in the low-voltage diode, but it becomes a problem when a product of length L_i and the resistivity of the pin diode's i-region are so high that it may become several hundred volts with the 5 kV diode, for instance. To turn on the pin diode, free electrons and holes that come in from the n-region and the p-region, respectively, must sufficiently fill the i-region, which occupies the major portion of that. And Eq. 2.17 is roughly estimated to the necessary on-time. Until then the pin diode

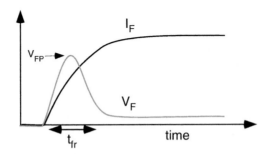

Figure 2.33 Forward recovery characteristics of a high-voltage diode.

is, so to speak, regarded nearly to be an insulator.

$$\frac{L_i}{v_d} = \frac{L_i}{\mu EF} \qquad (2.17)$$

L_i : i-region length $\qquad v_d$: Drift velocity

μ : Drift mobility $\qquad EF$: Electric field

Concerning the high-speed switching diode, this forward recovery characteristic rarely becomes a problem unless di/dt is particularly large.[60] However, as the influence of the superiority and inferiority of reverse recovery characteristics is serious, it is indispensable to design a suitable structure for high-speed operation.

FWDs for thyristors of several-hundred-hertz switching or transistors (BJTs) of 2–3 kHz switching need lifetime control[61] by introducing heavy metal atoms such as gold (Au) or platinum (Pt) atoms to them. If using a pin diode as the FWD for several-dozen-kilohertz or the rectifying diode for several-hundred-kilohertz high-voltage switching power supply, lifetime control, of course, is necessary. The method of lifetime control for them often employs irradiation by an electron beam (EB), proton ions, or helium ions

---

[60] It would be because L_i is designed short to lower the on-voltage and the accumulating charges in it simultaneously.

[61] Lifetime is the time constant to recombine or generate a hole and a free electron (refer to Section 2.5).

Moreover, the τ_g in Eq. 2.8 in Section 2.3.1 means the generation lifetime of them in a depletion layer and is the same as the recombination lifetime in principle.

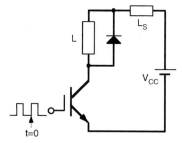

Figure 2.34 Measuring the circuit of reverse recovery characteristics.

because increasing the dose of heavy metal atoms increases the leakage current too much.[62]

Figure 2.34 shows an example of the measuring circuit of reverse recovery characteristics.[63] The value of L_S should be selected resembling the stray inductance of the assuming circuit, typically several hundred nanohenry, because an observation waveform changes by L_S, and it becomes an important parameter, particularly at evaluating the recovery destruction. On the other hand, since there is little contribution of the L value, several hundred microhenry to millihenry will be suitable.

By this circuit the circulating current flows in the diode after the first on-pulse and the recovery characteristics are measured at the time the second pulse is rising. Figure 2.35 is a typical result of the pin diode measurement. In the recovery operation V_{AK} decreases according to the current I_A attenuation and it becomes $V_{AK} < 0$ at the points a little after $I_A = 0$. The current-voltage $(I_A - V_{AK})$ characteristic of the diode is considerably complicated and is influenced by the following factors:

- Distributions of holes' and free electrons' densities in the on-state diode.

---

[62] Refer to Section 2.5.2.
[63] Such a chopper circuit is necessary because the traditional measuring circuit of the thyristor era cannot reproduce a high di/dt condition brought from IGBT operation. Using this circuit, one must be careful of the fact that observation waveforms greatly vary according to the characteristic of the driven IGBT. The driving IGBT must be selected to have a sufficient current margin to enable rapid control of measuring current.

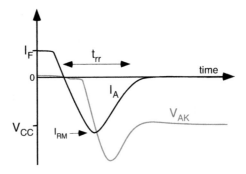

Figure 2.35 Reverse recovery waveform–I (basic characteristics).

(They depend on the on-state V_{AK}, impurity concentrations of p- and n-regions, a lifetime in the n$^-$-region, and so on.)
- Transient characteristic of the driving IGBT.
- Counter electromotive force by L_S.
- Counter electromotive force by L.
- Charge and discharge of the internal capacitor in the diode.
  (It can be ignored in the case of a pin diode.)

Figures 2.36 and 2.37 show electric charge distributions in the n$^-$-region that is in the reverse recovery process of a normal pin diode and a high-speed pin diode. First, in the on-state, the hole density n_h and the free electron density n_e are nearly equal in both cases, which are indicated as the waveforms of $I_A = I_F$. In the high-speed diode where the lifetime is shorter, the charge densities in the n$^-$-region are distributed like a pan-bottom shape, and their gross amount becomes small.[64] As large EF is necessary for flow of the same current, the on-voltage V_F increases.

When the current I_L, which flows through L, begins to diverge to the IGBT after the IGBT turns on in the second on-signal, the forward voltage V_F of the diode declines as an external force to make the current flow through the diode gets small. The accumulated amount of holes and free electrons in the diode are determined depending on the forward voltage V_F, then the holes and the free electrons are released according to the decreasing amount ΔV_F. This release causes the diode current I_A.

---

[64] Refer to Section 2.5.

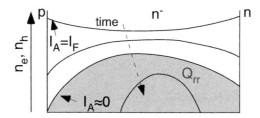

Figure 2.36 Temporal change of the charge distribution in a pin diode.

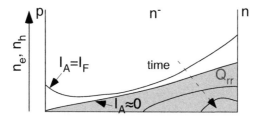

Figure 2.37 Temporal change of the charge distribution in a high-speed pin diode.

The rate of decline of charge densities is large mainly on both ends of the n^--region, which holds the voltage mostly. And the diode current becomes zero ($I_A = 0$) when an on-process of the IGBT proceeds and it takes all I_L. At this time, the charge density of the pn^- junction becomes approximately 0.[65] The amount of electric charges[66] that there are in the n^--region of this stage is approximately equivalent to the reverse recovery charge Q_{rr}. Furthermore, the charge density of the lowest portion at the on-state becomes a good indicator of Q_{rr} because the change of the charge distribution in the reverse recovery process is gentle at the pan-bottom.

When the discharge current of the diode decreases, V_{AK} becomes small, at $I_A \approx 0$, and is reversed later. Then the current begins to flow into L from a power supply, and the anode electric potential of the diode decreases by the counter electromotive force $-L(dI_L/dt)$.

---

[65] Because the important drift current comes to $J_{drift} = q(n_e + n_h)v = 0$, the charge density of the p-n^- junction must be ≈ 0.

[66] It means an amount of holes or free electrons. The recovery current is formed together with holes and free electrons moving at the same time.

Besides the fact that the increasing tendency of $|V_{AK}|$ causes a larger I_A, it controls the increments of I_L and $|V_{AK}|$. Nevertheless, in the situation where accumulated charges in the diode decrease, $|V_{AK}|$ and V_L suddenly increase and the IGBT shifts to normal on-operation.

For the diode to have this V_{AK}, a depletion layer where charges are extremely depleted and a high electric field exists in it is necessary. In the high-speed pin diode of Fig. 2.37, for instance, this area is formed after the holes and the free electrons around the p-n$^-$ junction side move to the anodal p-region side and the cathodal n-region side, respectively. The movement of these holes and free electrons is nothing but the current from the cathode to the anode in the diode. This is the reverse recovery current I_{rr}. And the sum of this I_{rr} and I_L becomes the I_{Ls} flowing through L_S.[67]

If there is no L_S the V_{AK} gets closer to $V_{CC} - V_{CE}$. And after reaching the peak, the recovery current I_{rr} would decrease slowly without depending on the V_{AK} directly.[68] In the presence of L_S, the countervoltage V_{L_S} is generated at the L_S by the decline of $I_{L_S}, = I_A + I_L$, when a declining rate of recovery current I_{rr} after the peak exceeds an increasing rate of I_L. The increase in this countervoltage V_{L_S} enlarges the reverse voltage V_{AK} of the diode and the adjacent L. As the V_{AK} exceeds the power supply voltage V_{CC}, a surge takes place. In addition, because the increase of V_{AK} promotes extending the speed of the depletion layer in the diode, the recovery current I_{rr} grows larger than the case without L_S, and it ceases in a shorter period. Such a situation corresponds to a hard recovery, mentioned later.

For an index of reverse recovery characteristics, the reverse recovery time t_{rr} had been used traditionally. However, the reverse recovery charge Q_{rr}, which is the integrated value of the recovery current I_{rr}, has been recently used because the typical recovery waveform as in Fig. 2.35 is not easily measured in the recent high-speed diode. Besides, the switching losses E_{OFF}, as well as E_{ON}, during chopper operation of Fig. 2.34 are sometimes used to evaluate the switching characteristics.

---

[67] It is also the collector current I_C of the IGBT.
[68] I_L would increase so that $dI_L/dt = -(V_{CC} - V_{CE})/L$ is valid.

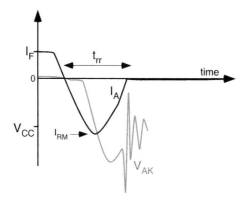

Figure 2.38 Recovery characteristics–II (oscillation).

The recovery phenomena can be classified into a hard recovery and a soft recovery according to the generated surge voltage. The former and the latter correspond, respectively, to the sharp and slow declining speeds of the recovery current I_{rr} after the peak.

In addition to these characteristics, an oscillation at the time of the recovery operation becomes a problem from the viewpoint of EMI noise reduction. Although internal charges decrease by an I_{rr} portion every second, a situation can occur that a large recovery current I_{rr} is flowing just before most of the internal charges disappear depending on the distribution and the amount of the accumulated charges in the diode. At that time, the diode behaves to change from a good conductor to something just like an insulator. In other words, as the recovery current disappears suddenly, a large surge is generated and an oscillation such as that in Fig. 2.38 happens. Such an oscillation is easy to happen in the hard-recovery diode.

In principle, it is desirable for the on-state charge distribution of a high-speed pin diode to be low in the anode p-region side and high in the cathode n-region side, as shown in Fig. 2.37. The former makes the recovery peak current I_{RM} small, and the latter is effective for the soft-recovery characteristics or the prevention of oscillation.

For the other important characteristic there is destruction at the reverse recovery operation, that is the recovery destruction. Owing to this destruction, it is considered that diodes rather than IGBTs

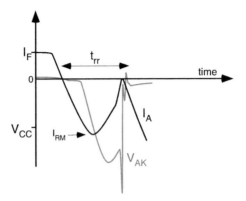

Figure 2.39 Recovery characteristics–III (recovery destruction).

cause inverter troubles in the market. The recovery destruction of this diode is one of the most difficult problems of the large-capacity semiconductor device, together with the destruction of high-voltage devices induced by a cosmic ray. Both are classified under the second breakdown.

As a recovery operation is a complicated phenomenon that a large number of factors affect, this type of destruction has not been observed enough yet. Nevertheless, it seems that there are two types of recovery destructions, one that occurs at approximately the maximum voltage V_{AK} and the other that happens after a terrible oscillation, as in Fig. 2.39. It is a common characteristic for all semiconductor devices to lose the holding voltage suddenly by the destruction.

2.4.3 Basic Operation of a pin Diode

The typical pin diode is schematically expressed in Fig. 2.26. For instance, the hole density n_h of each region is 10^{18} cm^{-3} for the p$^+$-region, 10^6 cm^{-3} for the n$^-$-region, and 10^2 cm^{-3} for the n$^+$-region. And at 300 K the free electron density n_e of each region is 1.2×10^{20} cm^{-6} divided by the hole density n_h.[69] This relation is called

---

[69]Actually, the n_e of the n-region almost becomes the phosphorus density and the n_h of the p-region almost becomes the boron density. The n_i is called the intrinsic carrier density.

the "law of mass action".[70]

$$n_h n_e = 1.16 \times 10^{20} \text{ cm}^{-6} \equiv n_i^2 \tag{2.18}$$

As mentioned in Section 2.2.3, at the border between the p- and n^--region an electric field toward the p-region from the n^--region exists and there is also an electric field toward the n^--region from the n-region at the n^--n boundary. Then, the internal electric potential of each region gets higher as the p-region $<$ the n^--region $<$ the n-region. Besides, if the hole was to conform to the Boltzmann distribution law, the next relations between the hole density n_{hp} in the p-region and the hole density n_{hn^-} in the n^--region is valid where an internal potential difference of p to n^- is expressed as V_{pn^-}.

$$\frac{n_{hp}}{n_{hn^-}} = \exp\left(\frac{-qV_{pn^-}}{kT}\right) \tag{2.19}$$

In contrast, for the free electron having an electric charge of $-q$, the density ratio of the same point becomes the reciprocal number in case of the hole.

$$\frac{n_{ep}}{n_{en^-}} = \exp\left(\frac{qV_{pn^-}}{kT}\right) \tag{2.20}$$

Multiplying Eqs. 2.19 and 2.20 by each side leads to Eq. 2.21. So it is apparent that the density product $n_h\, n_e$ of a hole and a free electron, respectively, is equal to the entire area if both holes and free electrons conform to the Boltzmann distribution law. The general relation (Eq. 2.22) that expresses the density products $n_h\, n_e$ in the neighboring regions A and B are equivalent is the original "law of mass action".[71]

$$\frac{n_{hp}}{n_{hn^-}} \frac{n_{ep}}{n_{en^-}} = 1 \tag{2.21}$$

$$(n_h n_e)_A = (n_h n_e)_B \tag{2.22}$$

By the way, the Boltzmann distribution law is valid in the thermal equilibrium condition.[72] In such a situation the charge carriers, holes, and free electrons are balanced between the neighboring

---

[70] Refer to Eq. 2.1, Section 2.2.3.

[71] In fact, the relation of Eq. 2.22 has a generality and can be induced from statistical mechanics or thermodynamics even if not assuming Boltzmann distribution law. In contrast, $n_h\, n_e = n_i^2$ (Eq. 2.18) is valid only in the thermal equilibrium state.

[72] Refer to Section 2.2.3.

areas having different inner voltages. The drift motion[73] caused by the inner electric field and the thermal diffusion motion caused by the density difference are in the opposite directions and cancel each other at the boundary. Then, if the balance is lost by applying the voltage between these areas, the thermal equilibrium would collapse because each charge carrier would move to one side.

However, in a situation in which the net movement of the charges is way smaller than the amount of drift motion or diffusion motion in thermal equilibrium, it can be considered that the Boltzmann distribution law is valid. Fortunately, the usual operating current density, below several hundred A/cm^2, of power devices fits into such a range.[74] Therefore, various joint portions of different areas, including pn junctions where an analysis seems difficult at a glance, can be interpreted easily for holes or free electrons to be conformed.[75]

In the above example, the behavior of the charged carriers which cross over the adjacent region is considered, but a situation in which an inner potential and the charge density change continually in the same region can be possible. The Boltzmann distribution law is valid and the corresponding relation to Eq. 2.19 or Eq. 2.20 is available between any two points about the hole or the free electron if their diffusion motion is substantially balanced with their drift motion. And their densities are uniquely determined by the inner potential.

From now, we think about a pin diode without electrodes. When a forward voltage is applied to it,[76] balances of holes and free electrons that are established at the p-n^- junction and the n^--n junction would collapse and the voltage differences of those junctions get smaller because the inner voltage in the p-region is originally lower than that in the n-region. Then, the diffusion

---

[73]The motion caused by an electric force qEF that acts on the electric charge q in the electric field.

[74]The device simulator shows that the amount of drift current and diffusion current flowing in the opposite direction at the junctions are tens of thousands of A/cm^2.

[75]Holes and free electrons behave like different kinds of ideal gasses, except for having an electric charge. Now, an example of a situation in which the Boltzmann distribution law cannot be applied is the depletion layer around the reverse-biased pn junction.

[76]It might be possible if many positive ions are irradiated into the p-region for instance.

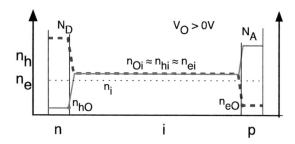

Figure 2.40 Carrier distributions of a pin diode (in the case of no electrode and $V_O = V_F > 0$ V).

motion of charged carriers becomes greater than the drift motion. Namely, the hole flows into the n⁻-region from the p-region at the p-n⁻ junction and the free electron moves in the reverse direction. Furthermore, the same principle acts on the n⁻-n junction and the hole moves into the n-region from the n⁻-region and the free electron also moves in the reverse direction.

Finally, the hole begins to move from the p-region to the n-region via the n⁻-region when the forward voltage V_F is applied and the free electron begins to trace the reverse course.[77] Then it shifts to a new stable state and the hole density n_h and free electron density n_e in the n-region, the n⁻-region, and the p-region become the new values, as shown in Fig. 2.40.[78] The applied outer voltage V_O is shared and applied to the p-n⁻ junction and the n⁻-n junction, where there was the original inner voltage difference.

Afterward, the hole density n_h and the free electron density n_e in the original n⁻-region are approximated to be equal to n_i (= 10^{10} cm⁻³) to simplify the explanation.[79] In Fig. 2.40 N_A and N_D express the hole density of the p-region and the free electron density of

---

[77] The device in which there are hole's and free electron's currents in such a way is called a "bipolar device."

[78] Holes and free electrons in this situation would be compared to positive and negative ions in the electrolytic tank during an electrolytic plating process. Metal ions begin to melt from the anode electrode into the plating liquid, and its metal ions and the negative ions in the electrolyte are distributed equally throughout the plating tank if the plating current is small.

[79] The n⁻-region is regarded as the intrinsic region, where $n_h = n_e$ to satisfy the charge neutrality.

the n-region, respectively.[80] And the applied forward voltage V_F is shared by V_{Fa} and V_{Fk} on the $p - i$ junction and on the $i - n$ junction, respectively, ($V_F = V_{Fa} + V_{Fk}$). It is because the potential difference cannot exist in a homogeneous region such as the i-region in a situation in which a charge carrier does not move continuously in it, that is, there is no current.

Then the free-electron density ratio $n_{en^-}/n_{en} = \exp(q V_{n^- n}/kT)$[81] across the n-n$^-$ junction is multiplied by the factor $\exp(q V_{Fk}/kT)$.[82] And at the n$^-$- p junction, the original free electron ratio $n_{ep}/n_{en^-} = \exp(q V_{pn^-}/kT)$ is multiplied by the factor $\exp(q V_{Fa}/kT)$.[83] Therefore, the original n_e at the p-side of the p-i junction, $n_{ep} = n_i^2/N_A$, is multiplied by the factor $\exp(q V_F/kT)$ and is expressed as n_{e0} in Eqs. 2.23 and 2.24.[84]

$$n_{e0} = N_D \exp\left[\frac{q(V_{n^- n} + V_{Fk}) + q(V_{pn^-} + V_{Fa})}{kT}\right]$$

$$= N_D \exp\left[\frac{q(V_{pn} + V_{Fk} + V_{Fa})}{kT}\right]$$

$$= n_{ep} \exp\left[\frac{q(V_{Fk} + V_{Fa})}{kT}\right] \tag{2.23}$$

$$n_{e0} = \frac{n_i^2}{N_A} \exp\left(\frac{q V_F}{kT}\right) \tag{2.24}$$

By the same way of thinking about the hole from the p-region to the n-region, the hole density of the n-side of the i-n junction, n_{h0}, can be confirmed to be expressed as Eq. 2.25.[85]

$$n_{h0} = n_{hn} \exp\left(\frac{q V_F}{kT}\right) = \frac{n_i^2}{N_D} \exp\left(\frac{q V_F}{kT}\right) \tag{2.25}$$

Now, at the i-side of p-i junction, the hole density n_h is multiplied by the factor $\exp(-V_{n^- p}/kT)$ and the free electron density n_e is

---

[80] N_A is approximately equal to the boron atom concentration in the p-region and N_D is approximately equal to the phosphorus atom concentration in the n-region.

[81] This equation might be induced easily by an analogy with Eq. 2.20.

[82] $\exp(q V_{n^- n}/kT) \ll 1$ and $\exp(q V_{Fk}/kT) > 1$. Be careful that there is no minus sign at the power of exp-term because a free electron is a negative charge.

[83] $\exp(q V_{pn^-}/kT) \ll 1$ and $\exp(q V_{Fa}/kT) > 1$.

[84] N_A is the original hole density of the p-region, $\approx$ boron atom concentration.

[85] N_D is the original free electron density of the n-region, $\approx$ phosphorus atom concentration.

multiplied by the factor $\exp(V_{n-p}/kT)$ to each value of the p-side. The density product $n_h n_e$ of the hole and the free electron becomes $n_i^2 \exp(qV_F/kT)$ equally in the i-region side and p-region side as in Eq. 2.26[86] after all.

$$n_{hi} n_{ei} = N_A n_{e0} = n_i^2 \exp\left(\frac{qV_F}{kT}\right) \tag{2.26}$$

The i-region where $n_{hi} = n_{ei}$ at $V_F = 0$ V must be $n'_{hi} = n'_{ei}$ even if $V_F > 0$ V.[87] In this way, the hole density n_h and the free electron density n_e in the i-region are both identified by Eq. 2.27. Also $V_{Fk} = V_F/2$ is induced because the n_e at the i-side of the n-i junction is increased from n_i at $V_F = 0$ V to $ni \exp(qV_F/2kT)$ by applying V_F. In other words, V_F is shared by halves with each junction, $V_{Fa} = V_{Fk} = V_F/2$.

$$n_{0i} = n_i \exp\left(\frac{qV_F}{2kT}\right) \tag{2.27}$$

After all, the n_h and n_e in the i-region are only determined by the applied voltage V_F.[88] And those values are uniform in the i-region and do not depend not only on the length of the i-region but also on the lengths and impurity concentrations of both the p-region and the n-region.

This mechanism that the density of holes and free electrons in the i-region is determined only by the outer voltage V_F is applied to an irregular diode, such as in Fig. 2.41a. Furthermore, it works when a narrow p^--region that contains a small amount of boron atoms is inserted in the n^--region, as shown in Fig. 2.41b. That is, the inner potentials of the n^--region on both sides of the p^--region are originally the same and the relative values would not be changed even if V_F was applied between A- and K-electrodes, because the voltage difference does not occur in the direction across the p^--region. Densities of holes and free electrons change on both sides of the p^--region depending on the internal potential difference between them,[89] but they do not change after passing through both borders.

---

[86] "The law of action" is also valid.

[87] "Electric charge neutrality" is one of the most rigid restrictions in nature.

[88] It is also affected by the energy gap E_g in the form of $n_i \propto \exp(-E_g/2kT)$.

[89] Holes and free electrons in the p^--region also obey the same principle and would increase when V_F is applied between A- and K-electrodes.

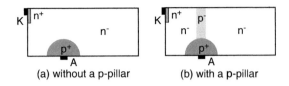

Figure 2.41 Substantially equivalent irregular pin diodes.

Furthermore, additional two-step preparations are necessary to explain the current flow in the diode. Those are (i) the contact phenomenon of a semiconductor and an electrode and (ii) the determining mechanism of the free electron's speed passing through the electrode.

Problem (i) is about the interconnecting mechanism between charged carriers in a pin diode and free electrons in its outside wire. Basically, even free electrons in the metal cannot be freely exchanged with the free electrons in a semiconductor. It is because a potential difference is induced if different materials touch each other, as mentioned in Section 2.2.4.[90] Then, for a free electron to come and go freely, first the n-type impurity, which is typically composed of phosphorus atoms, must be doped highly at the semiconductor surface and it is also well known empirically that it is effective to roughen the boundary between a semiconductor and a metal.[91]

When such an ideal contact is realized, what happens to the free electron density n_e of that point? It is believed that the n_e of the adjacent portion of the contact becomes approximately equal to the doped phosphorus density N_D and the law of mass action (Eq. 2.18) is also valid at that point, where the hole density n_h becomes n_i^2/N_D.[92]

Then how about the hole? Of course, there are no holes in metals. However, as in the case of the free electron, making the semiconductor surface p^+-rich and roughening the contact

---

[90] Such a problem occurs even between different metals, so it must be more serious when the case is between a metal and a semiconductor.
[91] For silicon, aluminum becomes a good electrode.
[92] It means that the thermal equilibrium state is established at the semiconductor adjacent to the contact boundary.

Diode Rectifiers | 79

boundary are known empirically to be very effective in getting the ideal connection.[93] In this case, it is also believed that the hole density n_h of the adjacent portion of the contact comes to be substantially equal to the doped boron density N_A and the free electron density n_e becomes $n_i^2/N_A \approx 0$.

In problem (ii)—free electrons speed passing through the electrode of pin diodes—the charged carrier distribution of Fig. 2.40 becomes the basis to consider. If adding the electrodes, the charge carrier distribution changes to what is shown in Fig. 2.42. When there is no electrode, the free electron density of the p-region is uniform, $n_{e0} = (n_i^2/N_A)\exp(qV_F/kT)$ (Eq. 2.24). However, at the p-region in Fig. 2.42, an incline of the free electron density, $\Delta n_e \approx n_{e0}$, occurs because $n_e \approx 0$ at the contact portion with an electrode, as mentioned in item (i).[94] When the thickness t_p of the p-region is less, free electrons move inline by diffusion mechanism exclusively. Then the product of $-q$ and the number of electrons that are moving per unit area and unit time becomes the current density J_e, (Eq. 2.28). D_e is the diffusion coefficient of the free electron.[95]

$$J_e = qD_e\frac{dn_e}{dx} \approx qD_e\frac{\Delta n_e}{t_p} \qquad (2.28)$$

Because free electrons pass through the device smoothly, the current density J_e in the p-region[96] becomes the J_e of the pin diode unless they disappear on the way.

The hole also moves from the p-region to the n-region through the i-region, lowering its density at each boundary part. In Fig. 2.40, with no electrodes, it becomes $n_{h0} = (n_i^2/N_{D0})\exp(qV_F/kT)$ in the

---

[93] Generally, the electrode metals suitable for the p-region are different from the ones for the n-region, but aluminum can be a very good electrode for both n- and p-type regions of silicon.

[94] In the practical current range, the deviation from the thermal equilibrium condition is so little that the law of mass action is still valid at the n$^-$-p junction. Then the free electron density n_e in the p-region near the n$^-$-p junction remains n_{e0}.

[95] D is related with the drift mobility μ as $qD = kT\mu$, which is called Einstein's relation.

[96] Generally, a drift current component might be added to this diffusion current. However, the former could be ignored because an EF is extremely small in the p-region where the hole quantity is so huge.

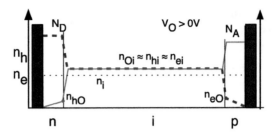

Figure 2.42 Charge carrier distribution of the on-state pin diode ($V_F > 0$ V, with electrodes).

whole n-region. However, in Fig. 2.42, with electrodes, an incline of the hole density, $\Delta n_h \approx n_{h0}$, occurs in the n-region because $n_h \approx 0$ at the contact portion with the electrode. When the thickness t_n of the n-region is less, the hole's diffusion current density J_h in it can be approximately expressed as Eq. 2.29.[97] And, as in the case of the free electron, this J_h value becomes the hole current density through the diode.

$$J_h = -qD_h \frac{dn_h}{dx} \approx qD_h \frac{\Delta n_h}{t_n} \qquad (2.29)$$

Many books make the mistake of saying that the recombination of holes and free electrons is important and indispensable for the operation of the diode or the bipolar transistor. The normal operation of pin diodes and bipolar transistors[98] can be explain without the recombination phenomenon[99] except the high-speed diodes, which are produced consciously to enhance the recombination rate. Although the recombination phenomenon exists to slight extent in a real device, the production technology that substantially removes its cause had been established by about 1970.

---

[97] A "-" mark disappears because variations of Δn_h and t_n are in the opposite direction each other. D_h is the diffusion coefficient of the hole, $qD_h = kT\mu_h$.

[98] Concerning unipolar devices such as the MOSFET, there is no need to consider recombination at all.

[99] The recombination phenomenon plays an important role as the reverse process to the collision ionization action considering destruction phenomena such as the huge current operation after breakdown or cosmic-ray-induced failures.

After all, the operating current of pin diodes can be expressed as Eq. 2.32 when the operating current is small.[100] Operating properties do not depend on the i-region length.[101]

$$J_F = J_e + J_h \approx q D_e \frac{\Delta n_e}{t_p} + q D_h \frac{\Delta n_h}{t_n} \tag{2.32}$$

Because Δn_e and Δn_h in these equations are nothing but n_{e0} (Eq. 2.24) and n_{h0} (Eq. 2.25), respectively, the small current operation of pin diodes can be expressed as Eq. 2.33.

$$J_F \approx q n_i^2 \left(\frac{D_e}{N_A t_p} + \frac{D_h}{N_D t_n} \right) \exp \left(\frac{q V_F}{kT} \right) \tag{2.33}$$

2.4.4 High-Voltage Large-Current Operation of a pin Diode

As a simulation example of a large-current operation, Fig. 2.43 shows distributions of current densities J_e and J_h, charge carrier densities n_e and n_h, and the EF in the pin diode with an n^--region of 100 μm[102]

---

[100] By the way, the current I_F flowing through the diode of Fig. 2.41 can be expressed as Eq. 2.30

$$I_F = I_e + I_h \approx q D_e \frac{\Delta n_e}{r} A_p + q D_h \frac{\Delta n_h}{t_n} A_n \tag{2.30}$$

In other words, the operating current I_F becomes small as the size of the n^+-region and p^+ portion of Fig. 2.41 is made small. Furthermore, if the ratio of the radius r of the p^+ portion to the thickness t_n of the n^+-region is increased, the free electron current I_e could be reduced largely compared to the hole current I_h. As such a situation is realized in the base emitter part of real bipolar transistors, the free electron current I_e of a whole current I_B can be ignored, as in Eq. 2.31. A_n should be regarded to be the emitter electrode width when the emitter region width t_n is as small as that in Fig. 2.41.

$$I_B \approx q D_h \frac{\Delta n_h}{t_n} A_n \tag{2.31}$$

[101] But they are restricted to the small current operation where the voltage in the i-region is negligible.

[102] In this calculation, uniform impurities densities and lengths of the n-region, i-region, and p-region are assumed to be 10^{18} cm^{-3}, 10 μm; 10^{14} cm^{-3}, 100 μm; and 10^{18} cm^{-3}, 10 μm, respectively, for simplification. And carrier-carrier scattering and the bandgap narrowing effect are included, but the recombination effect is not.

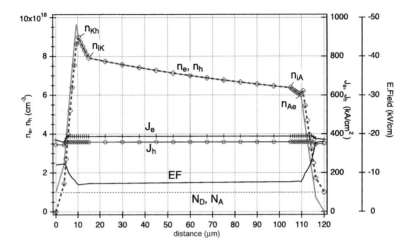

Figure 2.43 Distribution of J_h, J_e, n_h, n_e, and EF at $V_F = 100$ V. (n : n$^-$: p-region = 10 μm : 100 μm : 10 μm)

at $V_F = 100$ V. The 10 μm area on the left end is a cathode n-region and 10 μm area on the right end is an anode p-region.[103]

The on-current density at $V_{AK} = 100$ V indicates 747 kA/cm^2. At that time, charged carrier densities n_h and n_e in the i-region become almost 10^{18} cm^{-3},[104] where the EF is several thousands of kilovolts per centimeter.[105] The operating current J_F can be expressed as Eq. 2.34 approximately. This equation indicates the sum of two drift currents, one is the free electron current at the n$^+$ cathode electrode and the other is the hole current at the p$^+$ anode electrode. (N_D and N_A indicate impurities densities of the n$^+$ cathode and p$^+$ anode. μ_b and EF_a express the drift mobility and EF immediately adjacent to the electrodes.)

$$J_F \approx q\,N_D\,\mu_{b.e}\,EF_{a.K} + q\,N_A\,\mu_{b.h}\,EF_{a.A} \qquad (2.34)$$

---

[103] Impurities concentrations of the cathode n-region and the anode p-region, $N_D = N_A = 10^{18}$ cm^{-3}, are also indicated by a dotted line.

[104] Both carriers change linearly. Moreover, even in the p-region and n-region, n_h and n_e also change linearly when those region lengths are short, less than several micrometers.

[105] A strong negative EF originally existing at the n-i junction and the i-p junction almost disappears by $V_{AK} = 3$ V.

Furthermore, this equation can be reduced to Eq. 2.35 if $N_D \approx N_A \approx N$, as in Fig. 2.43. It is because the drift mobilities of the free electron in the n+-region and the hole in the p+-region become similar at high impurity concentrations[106] and the EF in the whole diode can be regarded to be nearly uniform. (L_{Di} indicates the whole diode's length.)

$$J_F \approx 2q \, N \, \mu_b \, E \, F_a \approx 2q \, N \, \mu_b \, \frac{V_{AK}}{L_{Di}} \tag{2.35}$$

It might be thought that such high voltage must not be applied to a diode. However, a thyristor must be forced in such a situation when it is misfired at the commutating timing, for instance. In addition, a GTO or a GCT must experience a similar situation at each turn-off or short-circuit operation.

2.5 Fast-Recovery Diode for a Typical Freewheeling Function

2.5.1 Need for First-Recovery Diodes

The high-voltage high-speed diode was used only for switching of less than 300 Hz in combination with a thyristor before the general-purpose inverter using a high-voltage BJT was manufactured.

By the way, it is necessary to connect an FWD to a transistor in reverse parallel, as in Fig. 1.6, because the transistor inverter is a voltage type. When the L-load such as a motor is driven by this inverter, a circulating current flows in the diode D_{i2}, which is connected serially to BJT1 in a series; BJT1 is turned off after conducting the main current.[107] Then, when BJT1 is turned on next, this circuit becomes the short-circuit condition through BJT1 and D_{i2}, during which the charge carriers accumulated inside the D_{i2} flow out constantly.[108]

---

[106] If the following are estimated: $N_D = N_A = 10^{18}$ cm^{-3}, $\mu_e \approx 230$ cm^2/Vs and $\mu_h \approx 150$ cm^2/Vs.

[107] The diving method of Fig. 1.6 is called "hard switching." In this circuit, the BJT cuts off the flowing current by itself forcibly and turns on bearing a high voltage.

[108] Namely, during the reverse recovery period t_{rr}. Refer to Section 2.4.2 for information on the reverse recovery operation.

When a normal pin diode was used as the FWD in the BJT inverter performing 2–3 kHz switching, it produced so much heat that it was destroyed immediately. Therefore, the reverse recovery period, or the recovery time t_{rr}, needed to be shortened as much as possible by shortening the lifetime of the FWD. So, it might be said that the high-speed diode truly appeared with the BJT inverter in 1981.

The lifetime control of the first-generation FWD that was combined with the BJT was performed by adding a diffusion process of gold atoms.[109] Although shortening of the recovery time could be possible by raising the gold diffusion temperature, it caused an increase in the leakage current, especially at high temperature, such as at 125°C. And too much shortening could bring about a thermal runaway. In addition, the on-voltage increased inevitably and there was the worry that gold atoms easily precipitate in the defective parts in a silicon crystal and produce an abnormal leakage current. It was a really difficult process.

Further shortening of the recovery time of the FWD was demanded when the switching frequency was raised to 10–20 kHz by using IGBTs. Because of the thermal runaway, the technique to raise the temperature of the gold diffusion reached the limit. Therefore, the platinum diffusion that caused a smaller a leak current than the gold diffusion had begun to be used.[110] However, enough trade-off between the short recovery time and the small on-voltage could still not be realized. Therefore, at the beginning of the 1990s, when a characteristic of the IGBT was improved to be competitive with the 600 V BJT, the characteristic lack of the diode became a more serious problem than the IGBT. It was a popular belief that there would be no room for improvement in a high-speed diode because of its simplicity.

However, it was improved by changing the anode p-region. At first the MPS structure mentioned in Section 2.4.1 was applied

---

[109]Just before the electrode formation process of the wafer, a thin gold layer was formed on the anodal side surface using a vapor deposition apparatus and then the wafer was heated up to about 900°C.

[110]It is caused by the difference of their intermediate-level energies E_t, that is $E_t \approx E_i - 0.14$ eV for platinum whereas $E_t \approx E_i$ for gold. The leakage current reduced to $\exp(-0.14/kT) \approx 1/220$ by changing from gold to platinum.

[9].[111] Furthermore, it could be improved only by making the anode p-region shallow and reducing its impurities [10] than has been the norm. Such a simple solution would seem to be left because nobody apparently had considered the essence of the diode operation.[112]

After that, it was submitted in 1996 that "the basic of the pin-structured power device is to do the charge carriers distribution low and flat at the on-operation by reducing the impurities doses of both p-region and n-region and also shortening the depths of them at the same time, and increasing the lifetime of the n^--region besides" [11]. It was about 2010 that the concept was understood clearly, as mentioned in Section 2.4.3.

Since the essence of shortening the reverse recovery time t_{rr} of diodes is making the quantity of internal accumulated electric charges Q_{rr} during the on-operation small, as shown in Fig. 2.37, it brings about an increase in the on-voltage V_F inevitably, as mentioned later. To reduce the increment of the on-voltage the length of the n^--region has been minimized as far as possible, keeping sufficient withstanding voltage.[113] Besides, a heavy metal, which was detested the most in a semiconductor process, must be attached to the wafer to intentionally introduce defects into a silicon crystal.

Furthermore, the internal charged carriers Q_{rr} must not only be small but also disappear gently in the last step of the off-process. Otherwise, the soft-recovery properties, mentioned in Section 2.4.2, are not provided. Soft-recovery properties are required from the aspect of EMI measures. It may be said that a high-speed diode is made under more severe restrictions than an IGBT.[114]

---

[111] However, the developers called it soft-and-fast-recovery diode (SFD). The concrete structure and the operating principle became clear long after it was developed.

[112] Even the policy of making the anode p-region shallow and lowering its impurities was not the result of theoretical consideration but the empirical result suggested with a know-how that the switching time of the BJT had been shortened by lowering its base impurity dose (in the case of Mitsubishi Electric Corp.).

[113] Shortening the n^--region can reduce both the quantity of the internal accumulated charged carriers and the on-voltage, but the destruction endurance property decreases necessarily. The destruction endurance property that is difficult to evaluate quantitatively would be apt to be put off as a design item.

[114] The IGBTs in recent mainstream have thin p-collector regions with low impurity doses and do not use lifetime control. However, for the high-speed diode, having

Moreover, an irradiation method of a charged beam of, for example, electrons e^-, protons H^+, or helium ions He^+ had been used as lifetime control also, as is mentioned later, in Section 2.5.3.

> After having irradiated them, an annealing process is performed.[115] This method has better uniformity and reproducibility than the traditional gold diffusion or the platinum diffusion. Although the irradiated electrons penetrate the chip, H^+ particles or He^+ ions in particular stop on the way and produce many defects in their neighborhood. This characteristic brings the big advantage that can control the lifetime at the narrow range of the depth direction in a chip. Nevertheless, a much expensive accelerator is needed for these irradiations.[116]

Although the diffusion wafer has been used for the high-speed diode traditionally because it is low priced and easy-to-get soft-recovery properties, the epitaxial wafer, which can meet detailed specifications, has begun to be used.[117] There are still many technical development factors on the high-speed diode at present, and the improvement demand is stronger than that for the IGBT.

By the way, a SBD does not have a recovery phenomenon theoretically. It is because the SBD works like a resistance in the on-state and charged carriers do not accumulate. However, the silicon SBD has been produced only below about 200 V.[118] Therefore, a SBD using SiC that would be usable even in a several-thousand-volt operation is looked forward to.[119]

---

only such structures of the p-anode region and the n-cathode region is insufficient and lifetime control is necessary in addition.

[115] Crystal defects created by the collision of high-energy particles are restored partially at about 400°C.

[116] Some power device makers entrust a supplier, whose main profession is supplying medical isotopes, with irradiation, but some makers have their own equipment. EB irradiation is much easier and can be entrusted to some suppliers who irradiate it into some plants or plastic films as their main profession.

[117] Refer to Section 2.7.1 for the diffusion wafer and the epitaxial wafer.

[118] The on-voltage of the higher-voltage SBD rises inevitably, and there are additional problems, for example, the leakage current is large and the operating speed gets slowed because of the appearance of holes when the impurity dose of the n^--region is brought low for a high withstanding voltage.

[119] 600 V and 1200 V products are commercially available from several companies now.

2.5.2 Effect of Lifetime Control

In Sections 2.2.3 and 2.2.4, it was mentioned that hole or the free electron could be considered to behave, or react, like an ideal gas atom in a semiconductor device. On this point of view, the lifetime could be interpreted as the time interval for the hole and the free electron to react.

In other words, if Object A collides with Object B, its reaction velocity R_A can be expressed as Eq. 2.36 or Eq. 2.37.[120] τ_A, which is usually called a lifetime, can be expressed as $\tau_A \equiv 1/(\sigma_B\, n_B\, v_{AB})$ generally, and it means an average interval of the reaction, or collision, that Object A causes for Object B.[121]

$$R_A = \sigma_B\, n_B\, v_{AB}\, n_A \qquad (2.36)$$

$$R_A = \frac{n_A}{\tau_A} \qquad (2.37)$$

$$\tau_A \equiv \frac{1}{\sigma_B n_B v_{AB}} \qquad (2.38)$$

τ_A : Average interval of the collision between A and B

σ_B : Cross section of B

n_A, n_B : Densities of A and B

v_{AB} : Relative velocity between A and B

By the way, the operation of a semiconductor device can be expressed in five basic equations, and the lifetime is related to Eqs. 2.39 and 2.40, which include the recombination velocity R.[122]

---

[120] In the case of chemical reactions or nuclear reactions, this is known generally as the kinetic theory of molecules, which began with the consideration of the motion of gas molecules.

[121] The term "life" is appropriate for τ_A only when Object A becomes extinct on collision with something.

[122] Both equations are collectively called current continuity equations. Although the meaning of R in the basic equations is originally a parameter indicating the continuity of the hole current density J_h and the free electron current density J_e, $\nabla(J_h + J_e) = 0$, it has been called the recombination velocity because $R = -\partial n/\partial t$ is valid if $\nabla J = 0$. The pair generation velocity G is indicated as $G = -R$.

Other than these equations, there are Poisson equation (Eq. 2.72) and equations

$$-\nabla \vec{J}_h = qR + q\frac{\partial n_h}{\partial t} \tag{2.39}$$

$$\nabla \vec{J}_e = qR + q\frac{\partial n_e}{\partial t} \tag{2.40}$$

R : Recombination velocity per unit volume

Both these equations show that the hole current J_h and the free electron current J_e are complementary. For instance, in the case of a pin diode, the free electron current $J_e(x)$ increases by $qR(x)$ as much as the hole current $J_h(x)$ decreased by $qR(x)$ in a certain place x, as shown in Fig. 2.44.[123] Such distributions of J_h and J_e are realized by the drift flows, primarily, and the diffusion flows, additionally, of holes and electrons that are flowing in opposite direction in the i-domain.

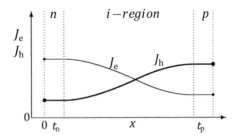

Figure 2.44 Distributions of J_h and J_e when recombination exists.

---

for the drift-diffusion transport model (Eqs. 2.41 and 2.42).

$$\epsilon_r \epsilon_0 \nabla^2 \psi = -q(n_h - n_e + N_D - N_A) \quad \text{(see Eq. 2.72 later)}$$

$$\vec{J}_h = -qD_h \nabla n_h - q\mu_h n_h \nabla\psi \tag{2.41}$$

$$\vec{J}_e = qD_e \nabla n_e - q\mu_e n_e \nabla\psi \tag{2.42}$$

The device simulator calculates an electric potential ψ and n_h and n_e so that these expressions are valid simultaneously at all points in the device. Other factors, for instance, J_h and J_e, can be reduced from them. The quasi-static simulation ignores $q(\partial n_h/\partial t)$ and $q(\partial n_e/\partial t)$ in Eqs. 2.39 and 2.40.

[123] Both J_h and J_e keep the same values if the recombination is ignored, namely $R = 0$.
In addition, Eqs. 2.39 and 2.40 indicate that a part of the hole current changes to the free electron current in a certain place. Getting that result, the recombination is not necessarily needed. For instance, the author thinks that the movement of an outermost shell orbit as the hole could be replaced by the movement of an outermost shell orbit as the free electron without the recombination.

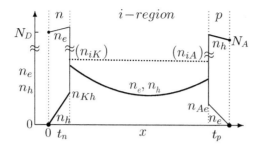

Figure 2.45 Distributions of n_h and n_e when recombination exists.

Of course, an electric charge neutrality condition must be satisfied in the i-region. As a result, it is in a situation that charged carrier densities n_h and n_e are distributed like a pan-bottom shape, as shown by the solid line in Fig. 2.45.

About these figures, the following can be said:

- The hole current $J_h(x)$ decreases toward the cathode side from the anode side, and the free electron current $J_e(x)$ increases in the same quantity inversely.
  Therefore, at the same outside voltage V_F, the hole current density $J_h(x)$ becomes smaller on the cathode side and the free electron current density $J_e(x)$ becomes smaller on the anode side than what it is in the case of no recombination.
  The current densities' distributions become the straight lines if the recombination velocity $R(x)$ is constant, irrespective of the place.[124]
- Since it is thought that the hole current J_h in the cathode n-region consists of a diffusion current mainly,[125] a smaller J_h means that the maximum hole density n_{Kh} in the n-region has gotten smaller than in the case of no recombination. Similarly, maximum free electron density n_{Ae} in the anode p-region has become small, too.

---

[124] It would be a proper approximation practically.
[125] In this book, pin diodes of the so-called transparent type, which have short cathode n- and anode p-regions have been discussed exclusively. In that case, the influence of recombination can be practically ignored in those regions.

- If n_{Kh} of the cathode region and n_{Ae} of the anode region become small, the charged carrier densities n_{iK} and n_{iA} at both ends of the i-region become small so the law of mass action should be satisfied.[126] In other words, the distributions of charged carrier densities n_h and n_e in the whole diode must be lower than the dotted line that links (n_{iA}) with (n_{iK}) in Fig. 2.45, which is for the case without recombination.
- To flow through the same current as J_F when there is no recombination, the EF in the i-region must become larger if charged carrier densities, n_h and n_e get smaller by the recombination process. Since it is nothing but that a voltage drop increases in the i-region, the forward voltage V_F gets larger inevitably when recombination exists.

After all, when there are recombinations between the hole and the free electron in the i-region, the charged carriers are distributed like a pan-bottom shape and an electric field $EF_i(x)$ appears in it. In the small current operation, when there is no recombination, the outside voltage V_F is only shared among the n-i junction and the i-p junction, namely $V_F = V_{ni} + V_{ip}$, but this assigned voltage $(V_{ni} + V_{ip})$ to these junctions decreases by an integrated value of inner electric field $EF_i(x)$ rather than V_F when recombination is strong. Then, because charged carrier densities n_h and n_e in the i-region are decided by the potential difference at each junction, these gross charges Q_{rr} decrease, unlike the case without recombination. The mechanism to improve the recovery properties by making the recombination velocity R large can be explained in this way.

2.5.3 Control Methods of the Lifetime

Lifetime control includes two methods, as described below.

- *Introducing the heavy metal atom*: Au diffusion or Pt diffusion into the silicon crystal has been the representative process,

---

[126] Equation 2.22, $(n_h n_e)_A = (n_h n_e)_B$, is the original form of the law of mass action and generally expresses the relation that the products of densities n_h and n_e of the hole and the free electron in areas A and B, which are next to each other as mentioned in Section 2.4.3, are equal.

and an inducing quantity is controlled by heat treatment, or annealing, after depositing the metal thinly. It is a technique that has been performed easily since the thyristor era, but it tends to be difficult to keep the uniformity and reproducibility. It is easy to be influenced by the quality of the semiconductor material, too.

- *Inducing crystal defects by irradiating high-energy particles*: As irradiation particles, electrons e^-, protons H^+, or He^+ ions have been used. In these cases, the energy and dose of the irradiation and the temperature and time of the annealing[127] after the irradiation are the control factors. Uniformity and reproducibility are good. It is a big advantage that there are few leakage currents in a high temperature.[128] Furthermore, since the depth of irradiation of heavy such as H^+ ions or He^+ ions can be controlled and, thus, they can be positioned in a narrow area, the carrier lifetime can be decreased locally in the device.[129] But special expensive equipment is needed to irradiate heavy particles.[130]

However, lifetime control is a really troublesome process because any process is far from the traditional semiconductor manufacturing method and its result can be easily influenced by the production history of the semiconductor material. In addition, a characteristic evaluation during the process is difficult.[131]

EB irradiation has been used now because productivity is good and the leakage current in a high temperature is small. In addition, the heavy particle irradiation method may also be used in combination with EB irradiation on devices for special usage such as those for very high-power conversion applications requiring a several thousands volt rating.

---

[127] After the irradiation, a low-temperature heat treatment of about 600 K must be performed to repair the damage in the crystal to some extent.

[128] Because of this problem, the gold diffusion was not used at first, and even the platinum diffusion later became a difficult method to adopt.

[129] It is a general trend that crystal defects are easy to be generated at the neighborhood where a high-energy particle stops and are scarcely generated on the way to it.

[130] A cyclotron is necessary to irradiate a H^+ ion or He^+ ion.

[131] The measurement of recovery properties is difficult, even if it is of the chip.

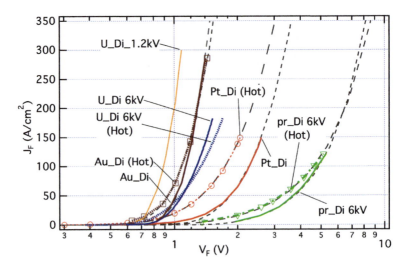

Figure 2.46 Observed/calculated J_F-V_F characteristics (with/without a label) (1.2 kV Au Di., 1.2 kV Pt Di., 6 kV H⁺ Di.). © [2006] IEEE. Reprinted with permission from Ref. [12].

By the way, the high-speed pin diode shows different J_F-V_F characteristics depending on the lifetime control method used for each. Naturally the device simulator should reproduce these characteristics, but it is common sense among power device developers that the device simulator cannot reproduce the phenomenon in which the lifetime participates. That is, if only using a Shockley–Read–Hall (SRH)-type recombination model, which is installed in a popular simulator,[132] even for the simplest pin diode it cannot reproduce the J_F-V_F characteristics that are shown in Fig. 2.46 for a large current area and in Fig. 2.47 for a small current area,[133]

---

[132] Such a simulator contains the Auger recombination also, but there was no indication that it brought any significant difference in the author's experience in power devices.

[133] "U_Di" means the usual diode without lifetime control. Samples were different from Fig. 2.46, but the manufacturing specifications of the Pt diode and the Au diode were the same. Besides, the calculated curves in these figures were the simulation results using the recombination models explained in Section 2.5.6.

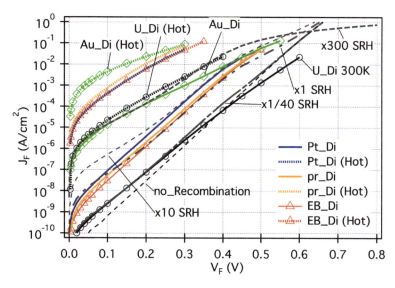

Figure 2.47 Calculated curves using SRH-recombination versus observed curves in a small J_F, V_F area. (Au, Pt, H$^+$, He$^+$, and normal diodes: 1.2 kV class). © [2006] IEEE. Reprinted with permission from Ref. [12].

and even the leakage J_R–V_R characteristics shown in Fig. 2.48.[134] If including temperature properties, the disagreement was much more.

2.5.4 Various Recombination Models

It has been said that direct recombination, or radiative recombination, which is expressed as $R_{\text{direct}} \approx \sigma_{\text{direct}} v_{\text{th}} n_h n_e$ (Eq. 2.43), is negligible because the capture cross section of reaction σ_{direct} is very small in an indirect semiconductor such as silicon. However, in a bipolar device, R_{direct} cannot be ignored because charged carrier densities n_h and n_e might be increased by several orders of magnitude in the i-region of a pin diode, for instance. In addition, it is natural that not only an SRH-type recombination but also a linear

---

[134]The manufacturing specifications of the Pt diode and the usual diode were the same as those in Fig. 2.47; in addition, $t_{n-} \approx 110$ μm.

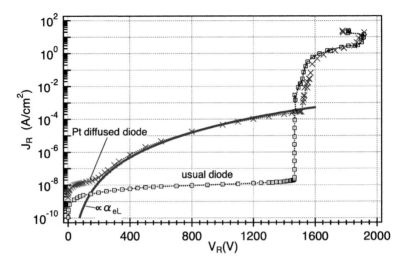

Figure 2.48 Examples of J_R-V_R curves of a Pt diode and a normal diode. © [2006] IEEE. Reprinted with permission from Ref. [12].

recombination, which is simply proportional to the charged carrier density, n_h or n_e, would exist.[135] Therefore, it would be proper to assume the following three kinds of recombination mechanisms:

- Direct recombination (radiative recombination)
  The thermal equilibrium state of the hole and the free electron in a semiconductor would seem to be caused by direct recombination. This idea could be understood by thinking that the expression of the mass action law $n_h\, n_e = n_i^2$ (Eq. 2.1) actually explains the reaction of a direct recombination and a generation between the hole and the free electron, namely $A \cdot B \Leftrightarrow AB$ (Eq. 2.3). The reaction velocity R_{direct} is expressed in Eq. 2.43.[136]

$$R_{\text{direct}} \approx \sigma_{\text{direct}} v_{\text{th}} n_h n_e \qquad (2.43)$$

---

[135] It was actually through the recombination mechanism that R. Hall thought to explain the observed J_F-V_F characteristics of the first pin diode expressed in Fig. 2.30.

[136] In the thermal equilibrium, R_{direct} is just compensated for the direct pair-generation velocity G_{direct}, namely $G_{\text{direct}} - R_{\text{direct}} = 0$.

Fast-Recovery Diode for a Typical Freewheeling Function | 95

- Linear recombination

  A reaction velocity $R_{\text{lin:e}}$ of the linear recombination that a free electron causes is expressed in Eq. 2.44, for instance,

  $$R_{\text{lin:e}} \approx \sigma_{\text{lin:e}} v_{\text{th}} N_t n_e = \frac{n_e}{\tau_{\text{lin:e}}} \tag{2.44}$$

- SRH-type recombination

  SRH-type recombination only acts on the deviated charged carriers[137] from the thermal equilibrium value. SRH-type recombination is complicated and explained next.

About the SRH-type recombination, R. Hall of GE manufactured a substantial pin diode[138] which was alloyed with indium and antimony, respectively, on both sides of the pure germanium piece in about 1950. He apparently had assumed that the on-current I_F flowed by direct recombination between the holes and the free electrons in the i-region and expected that I_F would be proportional to the squares of the charge carrier density, $n^2 \approx n_h \, n_e$. However, the observed I_F was simply proportional to n.[139] Then he appears to have also assumed that there would be a recombination that occurred through a specific level distributed in germanium. After the conference session reported in Ref. [13], Shockley developed a general recombination theory and derived Eq. 2.45 [14].[140]

$$R_{\text{srh}} \approx \frac{\sigma_e \sigma_h v_{\text{th}} N_t (n_e n_h - n_i^2)}{\sigma_h (n_h + n_i \exp^{\frac{E_i - E_t}{kT}}) + \sigma_e (n_e + n_i \exp^{\frac{E_t - E_i}{kT}})} \tag{2.45}$$

σ_e : Capture cross section of the intermediate level to the free electron

σ_h : Capture cross section of the intermediate level to the hole

N_t : Density of the intermediate level

---

[137] For instance, $\Delta n_e = n_e - N_D$, $\Delta n_h = n_h - n_i^2 / N_D$ in the n-region.

[138] Its structure is expressed in Fig. 2.30, in Section 2.4.1.

[139] He found that the on-current below 500 A/cm^2 was expressed as $J_F = q \ln \tau^{-1} [\exp(q V_F / 2kT) - 1]$, where τ was a constant.

[140] However, the recombination of Shockley and others (Eq. 2.45) is nothing but the mechanism that a charged minority carrier density that deviates little from the thermal equilibrium value returns to the natural value. Therefore, it could not essentially explain the large on-current of several hundred A/cm^2 that Hall observed. Nevertheless, the SRH-type recombination mechanism had widely spread, leaving somewhat unclear the original intention of Hall.

96 | *Basic Technologies of Major Power Devices*

In the SRH-type recombination model, an intermediate level E_t that is able to capture a charged carrier was assumed between the valence band E_v and the conduction band E_c.[141] The direct recombination of a hole and a free electron is the reaction that a free electron moves directly into the conduction band E_c from the valence band E_v. This transition rate could be accelerated by several orders of magnitude via a certain intermediate level E_t between E_v and E_c.

For instance, concerning the holes in the n-region, Eq. 2.45 can be transformed into Eq. 2.46.[142]

$$R_{srh.h} \approx \sigma_h v_{th} N_t \left(n_h - \frac{n_i^2}{N_D} \right) \approx \frac{n_h}{\tau_h} \tag{2.46}$$

: Recombination velocity of holes in the n-region

And concerning the free electrons in the n-region, Eq. 2.47 can be reduced. As a result, the free electron density n_e does not become smaller than the n-type impurity dose N_D.[143]

$$R_{srh.e} \approx \sigma_e v_{th} N_t \left(n_e - N_D \right) = \frac{n_e - N_D}{\tau_e} \tag{2.47}$$

: Recombination velocity of free electrons in the n-region

However, the exponential terms in the denominator of Eq. 2.45 cannot be ignored in a depletion layer where hardly any holes and free electrons exist, n_h, $n_e \ll n_i$. Now, concerning the major intermediate levels of the gold and platinum used for lifetime control, $\sigma_h \gg \sigma_e$, for both $E_t \approx E_i$ for gold and $E_t \approx E_i - 0.14$ eV for platinum. Then Eq. 2.45 can be approximated to Eq. 2.48 in the depletion layer where n_h, $n_e \ll n_i$, and the recombination velocity R_{srh} becomes a negative value,[144] $A \equiv \exp(\frac{E_i - E_t}{kT})$. The exponential

---

[141] Although E_v, E_c, and E_t originally meant the values of a valence band, a conduction band, and the Fermi level (or chemical potential), respectively, each is equated with its level itself afterward. The "intermediate level" is a very small abnormality portion of the crystal, namely the existence of a defect or an impurities atom, where a charge carrier has an energy between E_v and E_c when it exists there. It may be called the "trap level" or the "recombination center" from the functional standpoint.

[142] τ_h and τ_e are defined as $\tau_h \equiv 1/(\sigma_h v_{th} N_t)$, $\tau_e \equiv 1/(\sigma_e v_{th} N_t)$.

[143] This is the biggest characteristic distinguished from the linear recombination, mentioned later.

[144] On the fraction number, a bigger term in the denominator is more effective. Then, if σ_h and σ_e are at the same level, a tendency like the next equation appears

term of this expression brings the characteristic of the platinum-diffusion method of there being a much smaller leakage current than in the gold-diffusion method even the lifetime is shortened.

$$-R_{\text{srh}} = G_{\text{srh}} = \frac{\sigma_e \sigma_h V_{\text{th}} N_t n_i}{\sigma_h \exp^A + \sigma_e \exp^{-A}}$$

$$\approx \sigma_e V_{\text{th}} N_t n_i \exp\left(\frac{-|E_t - E_i|}{kT}\right) \tag{2.48}$$

(Pair generation velocity in a depletion region G_I)

In the SRH-type recombination, an inverse value of the recombination velocity $-R$ is nothing but the generation velocity G.[145] The generation velocity G_{srh} of the SRH-type recombination becomes maximum when the intermediate level E_t is in the center of an energy gap, namely $E_t = E_i$, and becomes small exponentially as E_t separates from E_i. Therefore, the intermediate level of $E_t \approx E_i$ can only contribute unless its capture cross section is huge, like a platinum atom. At that time, the generated velocity G_{srh} is expressed in Eq. 2.50 and is almost fixed with the larger value of τ_h or τ_e.

$$G_{\text{srh}} = -R_{\text{srh}} \approx \frac{n_i}{\tau_h + \tau_e} \tag{2.50}$$

(Pair generation velocity in a depletion region G_{II})

In addition, calling τ_h a lifetime of the hole is proper only for the phenomenon where a few holes become extinct in the n-region, as expressed in Eq. 2.46. And, similarly, τ_e should be used only for the phenomenon that a few free electrons would be distinguished in the p-region.

On the other hand, at the place where $n_h \gg n_i$ and $n_e \gg n_i$, as in the i-region of the pin diode at on-operation, Eq. 2.51 is reduced from Eq. 2.45. Originally, because Eq. 2.45 is the equation indicating the recombination of holes and free electrons, the amount

---

approximately.

$$\frac{1}{\sigma_h \exp^A + \sigma_e \exp^{-A}} \propto \exp^{-|A|} \tag{2.49}$$

[145] That is because the equation of R_{srh} includes both factors, the recombination and the generation. This relation is not valid for the general reaction velocity, which is expressed as $R_A = n_B/\tau_A$ (Eq. 2.37).

of variation Δn_h of the hole density and the amount of variation Δn_e of the free electron density are equal.[146]

$$R_{srh} \approx \frac{n_e n_h - n_i^2}{\tau_h n_e + \tau_e n_h} \qquad (2.51)$$

Furthermore, since $n_h \approx n_e$ in the real i-region in the on-operation, next Eq. 2.52 would be expected. Like Eq. 2.50 in a depletion layer, it is understood that the larger value of τ_h or τ_e has a meaning. That is, the lifetime of the SRH-type recombination model becomes $(\tau_h + \tau_e)$ even for the hole or the free electron.

$$R_{srh} \approx \frac{n_h}{\tau_h + \tau_e} \approx \frac{n_e}{\tau_h + \tau_e} \qquad (2.52)$$

After all, the SRH-type recombination was distinguished from the already known direct recombination in the following items:

- A charged carrier reacts through the intermediate level.
- It does not act on majority charged carriers when its density is less than impurities one.
- In an area where the reacting charged carriers are depleted by an electric field, it causes the generation reaction of a hole and a free electron.
- τ_e is usually different from τ_h,[147] but the recombination or generation velocity is determined with the longer τ substantially.

Many documents say that an SRH-type recombination model is enough for the recombination/generation model, or the lifetime model, of the charged carriers in the normal operation of the semiconductor devices.[148] However, the SRH-type recombination

---

[146] From this, it would be also clear that τ_h or τ_e is not worthy to be considered as the "existing time" of either the hole or the free electron, as mentioned at the beginning of Section 2.5.2.

[147] These are the time intervals when a hole or a free electron is captured by the intermediate-level E_t. Of course, the same number of both are recombined or generated in pairs.

By the way, in the direct recombination mechanism, because the hole and the free electron are recombined or generated in pairs, different lifetimes are unnecessary by all means.

[148] Although some documents add Auger recombination also, it has no influence when the charged carrier density is not as huge as about $n^3 \geq 10^{50}$ cm^{-3}.

is originally the phenomenon that is going to return a slight deviation of charged carriers from the thermal equilibrium state by the recombination or generation of them.[149] And the SRH-type recombination must be established only when the charged carrier density is considerably less than the intermediate-level density N_t because a charged carrier transits via the intermediate level. Therefore, it is not theoretically applicable to the situation where charged carrier densities are several orders more than the thermal equilibrium state, as in the i-region of the pin diode at on-operation.

2.5.5 Leakage Current Caused by Lifetime Killers

The on-voltage increases by all means when a lifetime killer (such as gold atoms, platinum atoms, or crystal defects) is introduced to make a high-speed diode, as mentioned in Section 2.5.2. And the tendency that a leakage current increases appears, too. The change of the control method of a lifetime described in Section 2.5.1 had been performed to avoid the increase of the leakage current at a high temperature mainly. Although the leakage current is the simplest characteristic of the semiconductor device, it might not be understood enough yet.

There are two kinds of leakage currents, the "generation current" and the "diffusion current." It is the former that increases the leakage current by introducing the lifetime killer. It has been said that only the generation current should be thought about with silicon devices because the latter is so small. However, with the power device, not only the generation current but also the diffusion current exists obviously.

2.5.5.1 Pair generation leakage current

When pair generation of a hole and a free electron occurs in a depletion layer, the hole and the free electron run for the opposite electrodes. Both together consist of a current between electrodes, which corresponds to the movement of an electric

---

[149]Because $R_{srh} = 0$ in the thermal equilibrium state, it is apparent that the SRH-type recombination never contributes anything to the formation of the thermal equilibrium state.

100 | Basic Technologies of Major Power Devices

charge of q.[150] As a mechanism of pair generation, there are the direct recombination model and SRH-type recombination model, mentioned in Section 2.5.4.

The direct recombination is the universal reaction taking place together with pair generation. Because it is thought that pair generation in a depletion layer would occur at the same velocity as the thermal equilibrium state, that would bring a leak current of Eq. 2.53.[151] (L means the length of the depletion layer)

$$J_{\text{R.direct}} = qG_{\text{direct}}L = qR_{\text{direct0}}L = q\sigma_{\text{direct}}v_{\text{th}}n_i^2 L \qquad (2.53)$$

Substituting $v_{\text{th}} = 10^7$ cm/s, $n_i^2 \approx 10^{20}$ cm^{-3}, $L = 100$ µm, and $\sigma_{\text{direct}} = 3 \times 10^{-20}$ cm^2.[152] to Eq. 2.53, $J_R \approx 5 \times 10^{-14}$ A/cm^2 is reduced. This J_R is way smaller than the actually observed $J_R \approx 10^{-8}$ A/cm^2 of the $L \approx 100$ µm pin diode.

On the other hand, the leakage current $J_{\text{R.srh}}$, which the SRH-type recombination brings about, was reduced to Eq. 2.54 using G_{srh} (Eq. 2.50). Then, if $\tau_h + \tau_e \approx 2$ ms, the observed $J_R \approx 10^{-8}$ A/cm^2 is reproduced. In this way, the SRH-type recombination assumes that an intermediate level brings a big effect. And the SRH-type recombination has become the established theory as the generation mechanism of the leakage currents in a depletion layer.

$$J_{\text{R.srh}} = qG_{\text{srh}}L \approx q\frac{n_i}{\tau_h + \tau_e}L \qquad (2.54)$$

However, it has the following problems which will be solved at the end of Section 2.5.6:

- The $\tau \approx 2$ ms that is expected by the observed leakage current is 2 orders larger than the expected lifetime $\tau \approx 10$ µs of the normal on-operation.
- It could be explained by the large separation between the intermediate-level E_t and the energy gap center E_i that the leakage current is not so changed after the irradiation of

---

[150]The current is generated at the moment when generated carriers begin to move. A leakage current is the extremely sensitive tool to detect pair generation phenomena.

[151] R_{direct0} in this equation is reduced from Eq. 2.43 when $n_h n_e = n_i^2$.

[152]This σ_{direct} was the estimated value by the author and it is also used in Table 2.1, which appears later.

electrons or protons.[153] However, if so, the influence on the on-operation by shortening the lifetime would not occur at all.

2.5.5.2 Diffusion leakage current

For instance, there are a few holes even in the n-region and an incline of the hole density n_h exists on the n-region side of the n-i boundary, as shown in Fig. 2.49a. The hole on the n-region side diffuses down at a certain velocity along this incline and is kept away from the n-region because of the built-in electric field EF_i existing at the n-i junction.[154] A similar phenomenon takes place with the free electron in the p-region, and together both diffusion currents become the leakage current (Eq. 2.55). And this leakage current increases from the relation of $L = \sqrt{D\tau}$ when the lifetime is shortened. Moreover, in both the pin diode and the p-n diode, the leakage current by this diffusion mechanism does not change only if the impurity densities of the p-region and the n-region are the same.

$$J_{R.diff} \approx q D_h \frac{n_{h0}}{L_h} + q D_e \frac{n_{e0}}{L_e} \tag{2.55}$$

n_{h0} : Hole density in the n-region $\approx \dfrac{n_i^2}{N_D}$

D_h : Diffusion coefficient of the hole in the n-region

L_h : Diffusion length of the hole in the n-region

n_{e0} : Free electron density in the p-region $\approx \dfrac{n_i^2}{N_A}$

D_e : Diffusion coefficient of the free electrons in the p-region

L_e : Diffusion length of the free electrons in the p-region

In addition, the situation of Fig. 2.49b might appear when the length of the n-region and the p-region t_n and t_p would become shorter than the diffusion length of each minority charged carrier,

---

[153] As shown in Eq. 2.48, an effect of the intermediate level whose E_t is far from E_i can be ignored because the capture cross section σ of such a level decreases exponentially with $|E_t - E_i|$. For instance, $E_t = E_v + 0.71$ eV for the intermediate level of an EB irradiation.

[154] Refer to Section 2.2.4.

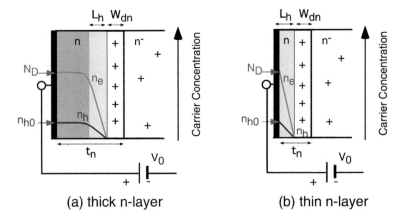

Figure 2.49 Generation mechanism of diffusion current $J_{R.diff}$.

L_h and L_e.[155] Therefore, the leakage current due to the diffusion currents at the junctions can be expressed as Eq. 2.56 and increases inversely proportional to t_n or t_p.

$$J_{R.diff} \geq qD_h \frac{n_{h0}}{t_n} + qD_e \frac{n_{e0}}{t_p} \quad (2.56)$$

t_n, t_p : Lengths of the n-region and p-region

Now, the charged carrier density product $n_{h0} n_{e0}$ of the hole and the free electron in the thermal equilibrium state becomes constant, as expressed in Eq. 2.57. So in the n-region of impurity density N_D, for instance, the thermal equilibrium density n_{h0} of the hole increases almost in proportion to $\exp(-E_g/kT)$, as shown in Eq. 2.58.

$$n_{h0} n_{e0} = n_i^2 \propto T^3 \exp\left(\frac{-E_g}{kT}\right) \quad (2.57)$$

$$n_{h0} \approx \frac{n_i^2}{N_D} \propto T^3 \exp\left(\frac{-E_g}{kT}\right) \quad (2.58)$$

---

[155] When ideal electrodes can be formed, the charged carrier densities n_h and n_e at the semiconductor surface in contact with each electrode become their thermal equilibrium densitiea n_{h0} and n_{e0}. (Refer to Section 2.4.3)

Therefore, concerning germanium, having a small energy gap, $E_g \approx 0.66$ eV, this diffusion current accounts for most of the leakage current.[156]

The diffusion leakage current of silicon devices is much smaller than that of germanium devices because $E_g \approx 1.1$ eV in silicon. Therefore, many books say that the diffusion leakage current does not need to be considered with silicon devices. However, concerning a high-voltage device that tends to use a lower impurity dose as a whole, it must be considered above dozens of degrees Celsius even if it is a silicon device[157] [15].

Furthermore, in recent years devices using extremely shallow n-regions or p-regions have become dominant. If the length of each region becomes very small, it is necessary for the diffusion leakage current to become much larger, as indicated in Eq. 2.56. In a structure such as that shown in Fig. 2.49b,[158] the diffusion leakage current is not affected by the lifetime value.

2.5.6 Interpretation of Observed J_F–V_F Characteristics

The serious problem that the device simulator cannot treat the lifetime appropriately, as mentioned in Section 2.5.3, would be solved by considering plural recombination mechanisms, including not only SRH-type recombination but also linear recombination and direct recombination[159] also, the author thinks.

---

[156]The usable temperature of germanium devices were confined to about $80°C$ because this diffusion leakage current was so big.

[157]It can be evaluated from the temperature dependency of an activation energy E_a on the leakage current J_R. The E_a of the generation leakage current is ≈ 0.6 eV, and the E_a of the diffusion leakage current becomes ≈ 1.2 eV because $n_i \propto \exp(-E_g/2kT)$ and $n_{h0} \propto n_{e0} \propto \exp(-E_g/kT)$. The typically observed E_a of the pin diode's J_R is $E_A \approx 0.6$ eV up to room temperature, but it increases and usually approaches 1 eV near about $125°C$.

[158]It is called a transparent structure because charged carriers go through the p-region or the n-region without attenuating.

[159]The SRH-type recombination would be dominant only in $J \leq 10$ mA/cm^2, and the linear recombination is superior to it. In addition, the direct recombination would be major over dozens of A/cm^2. In contrast, Auger recombination is not considered. It is because it will not affect substantially if it works in the charged carrier density over $n^3 \geq 10^{50}$ cm^{-3}, as described in the instruction manual of simulators.

Basic Technologies of Major Power Devices

Table 2.1 Optimized relative capture cross sections on Fig. 2.46

Diode type	Temp. (K)	$\times R_{direct}$	$\times R_{srh}$
6 kV (proton+EB)	300	50	4
	400	20	1.5
1.2 kV (Pt)	300	4000	70
	400	1000	17
1.2 kV (Au)	300	120	30
	400	80	17

Standard values: $\sigma_{d0} = 3 \times 20^{-20}$ cm^{-3} for direct recombination, corresponding to $\tau_{h0} = 3$ μs, $\tau_{e0} = 10$ μs for SRH-type recombination.

Actually, in the range of usual operation, a simulation using both a linear recombination and a direct recombination was able to reproduce the result of the observed J_F–V_F characteristics of various pin diodes almost exactly as shown in Fig. 2.46 [12].[160] Table 2.1 shows relative values of cross sections σ_{direct} and σ_{srh}, which were used in the calculations of the direct recombination and the SRH-type recombination at that time. In fact, the SRH-type recombination of this table expresses the linear recombination. In the current range of this figure, n_i can be ignored in comparison with n_h or n_e and there is no contradiction in that way because a linear recombination induces almost the same calculation result as the SRH-type recombination.[161]

The optimum SRH-type recombination velocities R_{srh} in Fig. 2.47, which indicate the J_F–V_F characteristics in a microcurrent range are about 1/10 of the platinum-diffused diode and about 10 times of the gold-diffused diode in comparison with the values in Table 2.1, which are optimized in the operation above about 1 A/cm^2. It seems natural to come to the conclusion that an SRH-type recombination is only effective in the microcurrent range[162] and

---

[160] Even with such a simple figure in it, adjusting the suitable parameters to accord in a two-dimensional curve was difficult. To reproduce the actual observed data, two kinds of recombination velocities of $\propto n^2$ type and $\propto n$ type were indispensable.

[161] Besides, it is also a reason for the interpretation that a specific constriction, which is not really observed, appears in the $J_F - V_F$ curve at about 1 A/cm^2 if using an SRH-type recombination.

[162] When gold atoms were doped, the typical situation appeared because $E_t \approx E_i$ was established.

above the miliampere range, a linear recombination is much more effective than an SRH-type recombination, the author thinks. This would be the solution of the problem in Section 2.5.5.1.

On the other hand, the ratio of the measured leakage currents of a platinum-diffusion diode and a gold-diffusion diode, whose production specifications were the same as that of the diodes in Fig. 2.48, was 10:500 at $VR \leq 200$ V. The fact that this ratio resembles the ratio of their very small forward currents (10:300) in Fig. 2.47 would be an example of the basic principle that the generation recombination speed $|R|$ is the same as the recombination speed $|G|$.[163]

2.6 Devices of the Thyristor Family

2.6.1 Thyristor

A thyristor has a structure that settles a p-region, which is called the gate, at the cathode side in the n^--region of the pin diode. When a positive voltage is applied to the p-gate with respect to the cathode, pin diode operation begins in a lengthwise direction near the gate.[164] Then, the diode operation spreads laterally, caused by the neighboring charged carriers in the i-region, and finally become a uniform operation in the whole device. In the on-state, both the internal charged carrier distributions and the electricity characteristic become the same as that of the pin diode without the p-gate region [16].

The basic operation of a thyristor can be expressed by the behavior of the hole and the free electron in one line from an anode to a cathode. Because it does not have a lateral structure basically, there is no factor to add as a variation. Besides, a model of Fig. 2.50 where a pnp transistor and an npn transistor were combined is traditionally employed to explain the thyristor operation. However, considering the simple operation, which is similar to that of the pin diode, the pin diode that is added to a thin p-region, as in Fig. 2.51, would be adequate as the basic structure of a thyristor. This

---

[163] It might be said for an SRH-form recombination at least.
[164] It is led by the operation of an npn transistor on the cathode side.

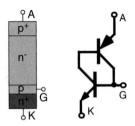

Figure 2.50 Structure and equivalent circuit of a thyristor.

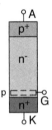

Figure 2.51 Basic structure of thyristor (II).

supporting p-region has the functions to turn on a diode and to hold the withstanding voltage. This is identical to the function of the gate of a thyratron.[165]

The thyristor is the first semiconductor device that can control a large amount of electricity, and the electric energy to be able to treat is the largest even now, that is, 12 kV 1.5 kA for a thyristor and 8 kV 3.5 kA for a light trigger thyristor. In addition, it was the only high-power switching semiconductor device until the BJT appeared. Until the mid-1980s, books on power electronics treated only thyristor circuits. Now, thyristors are used widely for direct current transmission, large-scale power conversion, high-power motor control, and so on. The 500 kV direct current transmission system, which uses a 250 kV switch constructed from 40 units of 8 kV 3.5 kA light trigger thyristors in a series for the +/− pole, is the greatest facility in Japan now.[166] Such a huge system is entirely in the hands of thyristors, but a thyristor has two big faults:

---

[165] Refer to Section 1.1, Chapter 1.
[166] It is used at 2.8 kA 1400 MW DC transmission at Kii-suido Channel.

- A commutation circuit is necessary to turn off the DC.
- The switching time is slow (less than about 300 Hz).

Not only is the commutation circuit expensive and bulky,[167] but also its electric losses are large and the commutation failure that sometimes happens by unknown causes is a serious problem.

2.6.2 GTO/GCT

In the use of thyristors it was known empirically that the main current of a small-capacity thyristor, of several amperes, could be cut off safely by pulling out a fairly large current from the gate. A GTO is a thyristor that enables this in a large current operation of more than several hundred amperes. The structure of a GTO is basically the same as that of a thyristor. Various steps had been adopted to get the ability to pull out a large gate current easily, such as dividing the cathode n-region into a lot of rectangles of narrow widths, surrounding them with the gate electrode, and raising the impurity dose of the gate p-region. Then a GTO could increase the effective reverse gate current and got a practical turn-off ability.

Nevertheless, a GTO is easily destroyed if stresses caused by the main current and the surge voltage, generated by a stray inductance L_S in the main circuit, at the interception operation are not relaxed by using a snubber capacitor that is arranged parallel to the GTO. A snubber shifts a phase of the current and the voltage of the GTO by taking over the current that should flow through the GTO to the parallel connected capacitor for a time, as shown in Figs. 2.52 and 2.53. However, the snubber circuit has problems such as the bulky sizes of the capacitor and resistor and loss of the whole snubber current on heating its resistor.[168]

GE developed a 300 V 7 A GTO in 1961 but could not commercialize it. However, Toshiba made a 600 V 200 A GTO experimentally in 1976 and put a 2.5 kV 800 A GTO to practical use in 1980. At that point a 4 kV 3 kA thyristor was manufactured. Then a 6 kV 6 kA GTO was manufactured in 1995. However, after that it

---

[167] It needs a large capacitor and an inductor in addition to an adjunctive thyristor and a diode.

[168] A capacitor of several mircofarad per 1 kA and a water-cooled resistor are needed.

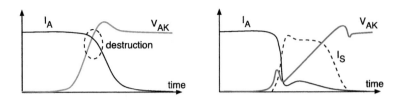

Figure 2.52 GTO turn-off waveforms (now snubber/with snubber).

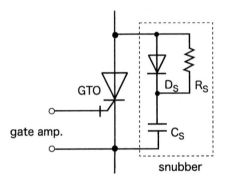

Figure 2.53 Typical snubber circuit.

was developed exclusively as a GCT. The representative uses of a GTO are the control system of the rolling mill of ironworks and the large-capacity power supply. It is used in the main power supply of the linear motorcar.

The GCT is the improved GTO that makes the snubber unnecessary.[169] The first improvement was to change a gate terminal from the conventional lead terminal (Fig. 2.54) to a whole circumference gate terminal, as shown in Figs. 2.55 and 2.56. Furthermore, the stray inductance of the gate circuit was reduced to the limit and a very low impedance power supply was used as the gate driver. Then it was able to throw all of main electric current I_K on a gate terminal from a cathodal terminal instantly. Then the whole main current

---

[169] A new driving system of the GTO named "hard-driven GTO," which simply uses a much larger gate current at turn-off operation than the traditional one paved the way for the GCT. That system realized a 100 MVA frequency converter for the electric railway by ABB Ltd. in 1996.

Devices of the Thyristor Family | 109

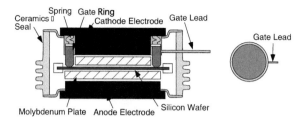

Figure 2.54 GTO package (cross-sectional view and partial plane view).

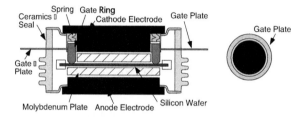

Figure 2.55 GCT package (cross-sectional view and partial plane view).

Figure 2.56 6 kV 6 kA GCT (the gate terminal is the disk with circular holes).

I_K was able to commute instantly into the gate terminal from the cathode terminal.

Figure 2.57 shows waveforms when the improved GCT is turned off by using the gate drive circuit of a very low impedance without a snubber. All the main current I_K that flows into the cathode from the anode commutates to the gate; therefore $I_G = I_A - I_K$ at any time. Then, the GCT is not destroyed even without a snubber [17]. This would be because the GCT does not operate as a thyristor but operates as a pip device during the period indicated by an arrow in Fig. 2.57.[170]

---

[170] Its operation is sometimes explained as a pnp transistor having a long base region, but there is no amplification function in the case of the high-voltage GTO.

110 | Basic Technologies of Major Power Devices

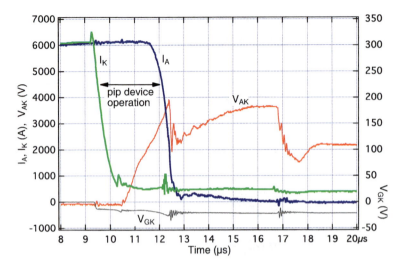

Figure 2.57 L-load turn-off waveforms of a GCT ($I_G = I_A - I_K$). Reprinted from Ref. [17] with permission from the Institute of Electrical Engineers of Japan.

Figure 2.58 shows distributions of the EF and charge carrier densities n_h and n_e of this period [17]. A neutral charge region consisting of accumulated holes and free electrons remains on the anode side of the n^--region, and an electric field exists on the cathode side. An anode current J_A consists of holes running through the EF region from the left-end portion of the neutral charge region toward the cathode. That is, the retreat of the neutral charge region causes J_A to flow toward the anode side accompanied by an increase

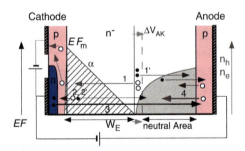

Figure 2.58 Distributions of electric field and charged carriers in the turn-off process. Reprinted from Ref. [17] with permission from the Institute of Electrical Engineers of Japan.

of V_{AK}. In the anode voltage V_{AK} where destruction becomes the problem, the hole's drift speed can be approximated to be constant, that is, the saturation velocity $v_s \approx 100$ km/s.[171] Then, the hole density n_h in the electric field region must be constant because an anode current J_A is expressed as Eq. 2.59. Besides, the inclination α of the EF is indicated by Eq. 2.60.[172]

$$J_A = q \, v_s n_h \tag{2.59}$$

$$\alpha \approx -\frac{q}{\epsilon_S} n_h = -\frac{J_A}{\epsilon_S v_s} \tag{2.60}$$

It should be noted that the main current of the GCT gets orders of magnitude larger than usual operation at the short-circuit operation and exceeds its turn-off ability. Therefore, inserting an anode reactance L_A in the main circuit as Fig. 1.4 is necessary to allow the GCT to turn off before I_A exceeds its ability.[173]

In the GTO era, it was desired that the gate driving current be small,[174] but on the contrary, ideal turn-off characteristics were achieved by increasing the reverse gate driving ability more than the main current. In addition, as the gate current increased but its turn-off time got 1 order of magnitude shorter than the conventional GTO's,[175] the gate driver's loss of GCTs was improved unpredictably. Therefore a GCT has a low on-voltage in comparison with a high-voltage IGBT while having the switching time at the same level.

In other words, the only words to describe the improving GCT was an overall reduction in the impedance of the device package and gate-driving circuit. Therefore, a series of GCT products nearly equal to the previous GTO product line had been manufactured in a few years. It was because a GCT could basically use the same semiconductor element, namely wafer, as a GTO. Furthermore, the integrated gate commutated turn-off thyristor (IGCT), which includes a gate drive system is manufactured to be usable without the delicate know-how of the gate drive method.

---

[171] Refer to Fig. 2.83.

[172] This situation is similar to the short-circuit operation of npn-type BJT in Section 2.7.3. The only difference is that the GCT has a pnp structure, not an npn one. (ϵ_S means the dielectric constant of silicon: $12 \cdot 8.85 \times 10^{-12}$ F/m)

[173] Refer to Fig. 1.4, in Chapter 1.

[174] Toshiba had succeeded in developing the GTO apparently by coming to the conclusion that there was no help for it without increasing the gate current to some extent at turn-off operation.

[175] The turn-off time of GCTs is similar to that of high-voltage IGBT's.

After all, it can be said that the thyristor needed 40 years to get its original potential finally as the GCT. In addition, as the thyristor and the GTO were most expected devices for power electronics until the 1980s, an energetic study was performed on them and the new basic structure of the high-voltage large-current device had been applied first. For instance, a cathode short structure, an anode short structure, various methods of lifetime control, a pin structure that is the original form of the famous field-stop structure of the IGBT, the structure including a reverse-conducting diode, and a module package were the techniques that were applied all ahead of BJTs and IGBTs.

2.7 Bipolar Junction Transistors

2.7.1 BJT Structures

The point contact transistor,[176] which was invented by J. Bardeen and W. H. Brattain, was released with a structure as shown in Fig. 2.59a[177] by Western Electric Company, a production section of AT&T, in 1950. However, as its operation was unstable, it declined when a junction-type transistor appeared in 1953—an alloy-

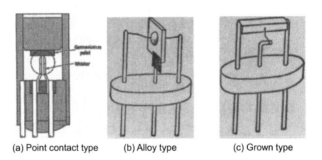

(a) Point contact type (b) Alloy type (c) Grown type

Figure 2.59 Assembly structures of early transistors.

---

[176] The "transistor" was named by W. Brattain on the advice of his coworker J. Pierce. J. Pierce meant the transresistance by an association from transconductance of a term used with a vacuum tube.

[177] An external case became the base terminal and had large holes to adjust contact wires. A user filled up the hole with resin after adjusting the whiskers.

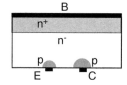

Figure 2.60 Expected structure of a point contact transistor.

type transistor from GE and RCA[178] and a grown-type transistor from WE.

The operating mechanism of the point contact transistor has not been well explained yet, but Fig. 2.60 seems adequate as the schematic diagram. The p-regions of the emitter and the collector are formed by melting an electrode metal or its containing impurities at the "forming process," which adds a pulse current between the collector and base terminals.[179] Only high-frequency characteristics were superior to an early junction-type transistor.

To make an alloy-type pnp transistor (Fig. 2.59b), for instance, small grains of indium of p-type impurities are pushed to both sides of a single-crystal thin chip of n-type germanium and heated in a high temperature, of 600°C, in hydrogen or an inert gas. In begins to melt at 160°C and dissolves Ge at 20% fusibility as the atomics ratio at 600°C. The Ge is precipitated from the In solution and recrystallized if the temperature is decreased slowly. Indium atoms included in this recrystallized Ge act as a hole-supplying source, namely the acceptor.[180] As the main process started with one heat treatment,[181] the alloy method was simple and easy, but the chip thickness was influenced by the history of the processing

---

[178] As for the patent application of alloy-type transistors, RCA Corp. was early by one day but the technical anticipation of GE was accepted after a dispute of seven years.

[179] This process was needed to get a good characteristic before using. Besides, it was left to a user. In addition, the forming of the emitter portion would happen naturally using a transistor.

[180] It becomes the structure on the left of Fig. 2.61. The structure on the right of Fig. 2.61, which is well known as the schematic diagram of transistors, is really proper for this alloy-type transistor.

[181] The connection with the base electrode was able to perform during the same heat treatment process, too.

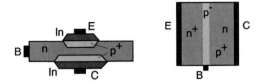

Figure 2.61 Structure and schematic diagrams of an alloy-type transistor.

temperature, the indium quantity, the crystalline axis,[182] and the crystal defect, then yield (the good product rate) was bad. As the interception frequency f_α where the output electricity became 1/2 was about 1 MHz, it was not usable as a radio frequency and was used as an audio frequency exclusively.

The crystal for a grown-type transistor was made of a crystal raised up slowly while spinning around the raising direction after soaking a seed crystal into dissolved germanium.[183] To make npn-type, at first a collector region was grown doping n-type impurities, antimony, and then a base layer[184] was grown doping p-type impurities, gallium, following it. An emitter region was grown doping n-type impurities again. After having sliced this crystal thinly, the surrounding portion of the junctions was cut down in a rectangle that was a few millimeters long and a few tenths of a millimeter square as a cross section. After having removed the surface processing damage by etching, both ends were soldered to lead wires of the collector and the emitter as shown in Fig. 2.59c.[185]

Wiring to the base region of about 20 μm width was done by electric welding. At the first welding terminal, a Ga containing gold wire of about 25 μm diameter was attached to touch the junction neighborhood sequentially and the change in the potential difference with the crystal was observed and the minimum point identified. When a pulse current was made to flow at that time, the gold wire welded to the Ge bar and a molten point spread out not only into the base region but also into the collector and emitter

---

[182] The (111) plane where a recrystallized face became flat was used.
[183] A germanium crystal of diameter ≈ 3 cm was got from Ge pellets of about 65 g.
[184] Its thickness was about 15–25 μm.
[185] As the solder adhesive, Sn-Sb was used for the n-type semiconductor and Sn-In was used for the p-type.

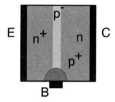

Figure 2.62 Schematic diagram of a grown-type transistor.

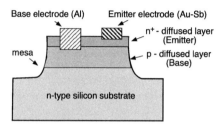

Figure 2.63 Original silicon high-voltage BJT for TV (double-diffusion type).

region. And after it got cold, Ga was left, and a p^+-region was formed. Therefore, it is appropriate that Fig. 2.62 refers to the grown-type transistor.

The silicon transistor had started in a mesa structure, as shown in Fig. 2.63, since about 1956. Then it accomplished great development by using a planar technology[186] invented in 1959 as its reliability and productivity had increased a lot. Although it had been already known that a silicon-oxide film masked the impurities diffusion, leaving the silicon-oxide film, which was formed at the first wafer process, until the last except an opening area for base diffusion is the planar structure.[187]

And a technique to build up the silicon crystal of low impurity concentration on the highly dosed large-diameter substrate appeared as shown in Fig. 2.64 in the mid-1960s. Combining this silicon wafer and the selective diffusion method using the mask of silicon-oxide film enabled the construction of the transistor

---

[186] US Patent US3025589: 1962, by J. Hoerni of Fairchild company.

[187] It was named considering that a chip surface looked like a plain whereas "mesa" means an isolated flat-topped hill with steep sides.

Figure 2.64 Epitaxial and substrate wafer.

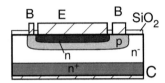

Figure 2.65 Si epitaxial planar-type transistor.

structure simply as in Fig. 2.65. This structure was called an epitaxial planar-type transistor. On the planar-type chip, p-n junctions were never exposed, being covered by an oxide film during any process, so the production yield in the manufacturing process was largely improved. Furthermore, it brought superior long-term reliability even if a molding resin sealing package was used. Therefore, mass production was enabled in both the wafer process and the assembling process and the product price was extremely reduced.

By the way, to obtain a higher withstanding voltage the resistivity of the region holding the voltage must be raised, that is, the base substrate for the alloy-type transistor and the collector region for the growth-type transistor. In those regions, a certain width is necessary for an electric field existence region, namely a depletion layer, to extend when a high voltage is applied. Then, the alloy-type transistor whose base width is essentially thin, less than about 20 μm, is not suitable for a high-voltage transistor. Besides, the grown-type transistor has enough collector thickness but is not able to treat high power because its collector region works as a resistance adversely.

However, the epitaxial transistor can design freely both the impurities density and the thickness of the collector active region.[188] For instance, if a thin and high-resistivity n^--layer was formed on the n^+-substrate of low resistivity, it would be balanced by a

---

[188] The original purpose of the epitaxial manufacturing method was to develop a high-frequency transistor for military purposes.

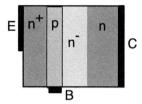

Figure 2.66 Schematic diagram of an epitaxial transistor.

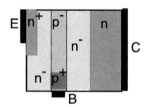

Figure 2.67 General schematic diagram of a bipolar junction transistor.

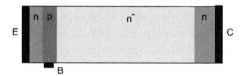

Figure 2.68 Schematic diagram of a high-voltage transistor.

good withstanding voltage, a high-power operation ability, and high-speed operation. It may be said that the low-voltage transistor was completed in an epitaxial planar structure after all. Although an epitaxial transistor can be expressed as in Fig. 2.66 in an extremely simple schematic diagram, Fig. 2.67 could be the more precise and universal diagram, because both the base p-region and the emitter n-region are formed by impurities diffusion from the surface.[189]

Practical use of the high transistor substantially began with the horizontal deflection circuit use of the TV. As it treats a voltage of about 10 kV, a high-voltage transistor of about 1500 V is desirable. For maintaining such a high voltage, using silicon is inevitable and the resistivity of the collector n^--region becomes dozens of Ωcm and more than dozens of μm are necessary for the length. The schematic

---

[189] It would be much more reasonable to think that the characteristic of the pin diode does not depend on its i-region length, as mentioned in Section 2.4.3.

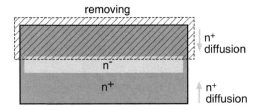

Figure 2.69 Wafer substrate of the triple-diffusion method.

diagram of the high-voltage transistor is shown in Fig. 2.68. In the beginning it was made with a method to thicken the n⁻-region of the epitaxial transistor, but the epitaxial growth process was expensive.[190] Then those transistors changed into a triple-diffusion type from the silicon-epitaxial-mesa type from 1970 through 1972, when color televisions spread rapidly in the US and Japan. The withstanding voltage BV_{CBO} between the collector and the base was 1500–1700 V, and the current capacity was about 5 A.[191]

Triple diffusion means that a collector region is formed by diffusion in addition to the base and emitter regions. By this method, n^+ diffusion is performed to a depth of more than 100 µm across the entire surface of an n⁻ wafer, as shown in Fig. 2.69, at the beginning.[192] Next, the wafer is polished, removing the n^+-diffusion portion from one side and the n⁻-region of a necessary thickness is left, depending on what the withstanding voltage should be. To form deep diffusion, the wafer must be kept at a high temperature, about 1250°C, for many days, and the precision abrasion technology is necessary, but the wafer price does not change even if the withstanding voltage becomes higher.[193] As the n^+ impurities levels

---

[190] In the first place, the manufacturing equipment was very expensive and it took time and money in proportion to the n⁻-region length.

[191] The withstanding voltage BV_{CEO} between the collector and the emitter was 600–700 V. They were sealed in a metal can called a TO3-type package.

[192] The n^+ portion accounting for most of the thickness of the wafer forms very thick because it is necessary to keep the machine strength of the wafer.

[193] It rather becomes cheaper as much as n^+ diffusion gets shallower. Whereas the epitaxial process suggests refinement or elegance, this collector diffusion is an unsophisticated and unrefined process. However, the Japanese high-voltage transistor for the TV could overwhelm the United States by this technique, the author thinks.

formed by diffusion are several orders lower than the n^+ substrate of the epitaxial transistor, it is inferior in terms of large-current ability compared to an epitaxial transistor. However, the advantage of price far exceeds this disadvantage.

On the other hand, some transistor makers out of TV business focused on the electronic ignition, which replaced the mechanical contact point of the ignition system of the automobile engine with a transistor, that had $BV_{CEO(sus)} \approx 300$ V and was of several amperes. Then, withstanding property for the stress in the turn-off process of an ignition coil was newly involved. Comparing this usage, the stress in the horizontal deflection circuit of the TV was much smaller because the transistor's operation in the application was not a switching operation but an inductive-capacitive (L-C) resonant load-driving operation. The measure that further high voltage was not added to the BJT was taken by connecting a diode whose withstanding voltage was lower than that of the BJT in reversely parallel to the collector emitter of the BJT.[194] It was because a diode had much a higher withstanding property than the BJT. In addition, the Darlington connection of Fig. 2.70 was adopted as the countermeasure against the problem that the current amplification factor h_{FE} decreased necessarily when the n^--region was thickened to get a high voltage of ≈ 300 V.

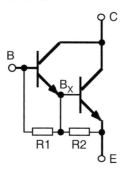

Figure 2.70 Darlington connection of the BJT.

Separately from these consumer uses, there was demand for military and industrial uses, such as the gun turret drive in the

---

[194] Moreover, a diode might connect in reversely parallel to the collector base of the BJT. Then the size of the diode could be reduced to $1/h_{FE}$ and the diode might be built into the BJT chip.

120 | *Basic Technologies of Major Power Devices*

tank, which is several hundred volts and about 200 A, and those were made to order using the same wafer-processing and packaging technologies as those in the thyristor and the GTO. In addition, there was the movement to replace the electron tube for the output stage of the medium-wave broadcasting with a transistor.[195]

In those days, the inverter device using the transistor (BJT) that had the problem of the second breakdown had existed. However, it was special equipment that had to be used unavoidably when the thyristor with a commutated circuit could not be used.[196] The power electronics apparatuses such as the inverter were built to order, and large-capacity thyristors, GTOs, and BJTs used for them were also made to order.

This was the situation around the power transistor in about 1973 when the Newell speech was made, as mentioned in Section 1.2, Chapter 1. After that, power transistors grew rapidly in a few years as noted in Section 1.3, Chapter 1. However, its growth was achieved by developing a transistor of several amperes for automotive ignitors or the switching power supply and not by the study that aimed at the large-capacity transistor for industrial use, as Newell expected.

The GTO was expected to have been a high-voltage power device exclusively, and the BJT was not expected so much until a transistor module was put to practical use. It was considered that the BJT would be destroyed easily by the second breakdown when it was used in a high-voltage situation. This second breakdown occurred when the main voltage V_{CE} rose by the reverse electromotive force of the L-load before the main current J_C decreased sufficiently at the turn-off of the BJT. In other words, an ability to hold a high voltage while a severe current flowed in the BJT was required.[197]

However, Fuji Electric manufactured a $V_{CE(sus)} = 450$ V[198] 50 A BJT that could operate the inverter[199] [18] in 1978. Furthermore,

---

[195] It was 300 V and 5 A, but high-speed switching was demanded. Afterward, the semiconductor transmitter was realized using the SIT or the MOSFET.

[196] It needed the user's effort to change the design and adjust the snubber with every production group of the BJT.

[197] Refer to Section 2.3.3.

[198] $V_{CE(sus)}$ was measured at a low current, about 1 A. For $V_{CE(sus)}$, refer to Fig. 2.28.

[199] The second breakdown was overcome by an appropriate design of the base drive circuit and a simple snubber circuit

Bipolar Junction Transistors | **121**

in 1980, Mitsubishi Electric began commercial production of power transistor modules integrating BJT chips that did not require snubbers in their applications. Nevertheless, it was destroyed easily in L-load operation with the higher-voltage BJT even if its withstanding voltage was raised.[200] This problem was solved by rejecting the slow operating portions of the BJT,[201] as mentioned later.

After that, when developing the BJT for the AC 400 V line, $V_{CE(sus)} = 900$ V, a problem appeared: the endurance voltage for the short-circuit test was lower than the desired one. It was a severe examination to turn off safely after 20 μs operation keeping the power supply voltage between the collector emitter directly at the usual base current, and then collector current reached approximately three times the rating current.[202] When the applied voltage V_{CE} was increased gradually, the BJT was destroyed suddenly after 2–3 μs[203] though its temperature did not increase a lot. Then, it was revealed that a new second breakdown existed. The BJT was destroyed instantly when the collector voltage and the current density product $V_{CE} J_C$ reached a critical value, ~ 200 kW/cm^2,[204] as shown in Fig. 2.71 [3]. This problem was settled by lowering the aim voltage on the user's recognizing this destruction mechanism and its quantitative principled limit.[205]

In addition, about the second breakdown at the turning off of L-load operation, it became clear that the BJT was not destroyed at a current density less than the specific collector current density, $J_0 \equiv q N_{Dn-} v_s$,[206] and the operating voltage range spread by increasing the reverse-base current against common sense, in 1985, as shown

---

[200]That is, the collector n$^-$-region was extended and its impurities density was lowered.

[201]This could be done easily by eliminating the corresponding parts from the photolithography mask pattern, especially the emitter mask.

[202]The BJT developer recognized that this endurance property was required to use it for an inverter for the first time.

[203]At this time the collector current rose and became the maximum.

[204]The value was arrived at by dividing the collector current I_C by the emitter area. If divided by the chip area it decreased to ≈ 100 kW/cm^2.

[205]Refer to Section 2.7.3.

[206]N_{Dn-} is the impurity concentration of n$^-$-region and v_s is the saturation velocity. Refer to the explanation of Table 2.2 in Section 2.7.3.

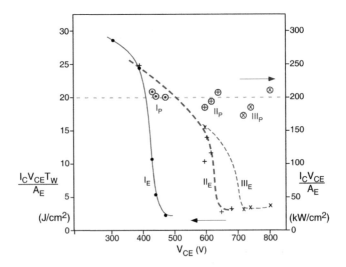

Figure 2.71 Short-circuit destruction energy and power versus V_{CE} (measured values on BJTs of three different n⁻ lengths). Reprinted from Ref. [3] with permission from the Institute of Electrical Engineers of Japan.

in Fig. 2.72 [20].[207] The small destruction ability of the BJT in L-load operation was caused by the extra or parasitic structure in a transistor chip where turning-off operation was slower than the main portion. It became clear that the transistor works normally until just before destruction and even the sudden drop of V_{CE}, which was judged conventionally as destruction, indicates merely the change of its operating mode in a severe current and a high-voltage operation.

After Mitsubishi Electric manufactured a 1000 V 300 A transistor module in 1982 in this way, inverters using BJTs spread rapidly because of the robustness of the BJT chip and the ease of module packaging [3].[208] After this, the BJT chip did not

---

[207] The operating points, namely trace, of $I_C \leq 10$ A show that the "sustain operation" holds a certain voltage that is determined exclusively by the current value flowing in the BJT. It is nothing but a situation in which a large current flows while holding the net potential withstanding voltage of the semiconductor device by itself.

[208] The long-term reliability of withstanding voltage properties became a serious problem for module sealing of a planar-type high-voltage BJT, but it was solved by using the structure shown in Fig. 2.25. Planar-type devices of more than 6 kV are manufactured now. The main high-voltage devices such as MOSFETs and IGBTs,

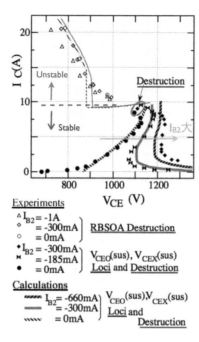

Figure 2.72 Measured and simulated operation limits. 1,000 V 10 A a single BJT (at 25°C). © [1992] IEEE. Reprinted with permission from Ref. [19].

change in its basics until its production stopped in the beginning of 2000s.[209]

Schematic diagrams of BJTs from Fig. 2.73 to Fig. 2.76 show the improving history of the BJT's structure. These diagrams clearly show that the emitter region, which was far from the base electrode and operated slower than other portions, had been removed intentionally. Figure 2.77 shows the aluminum electrode pattern of a 1000 V 75 A BJT chip manufactured in 1982. A large number of rectangular small emitter regions were formed dispersing under each stripe of the second and third stages of the Darlington-type transistor construction.

---

except large thyristors and diodes, use this method to reduce a potential difference between the top and bottom of an oxide film on the chip surface.

[209] As an exception, steps of the Darlington connection had been added by a new design concept in around 1990 to increase its h_{FE} characteristic for reducing the base drive current and, thus, enabling reduction of the base drive unit size.

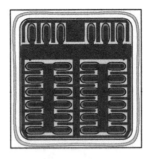

Figure 2.73 Plane view of the BJT for inverter-I (with electrodes).

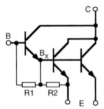

Figure 2.74 Two-stage Darlington BJT construction.

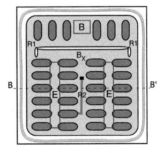

Figure 2.75 Plane view of the BJT for inverter-II (without electrodes).

2.7.2 Basic Operation of the BJT

An epitaxial planar-type BJT is expressed in Fig. 2.67. It might expand more generally to Fig. 2.78 for the npn BJT. This structure has the following characteristics:

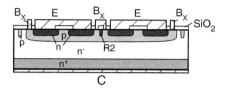

Figure 2.76 Cross-sectional view of the BJT for inverter (B-B' portion).

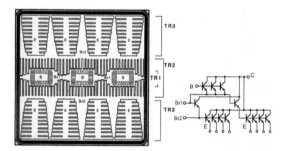

Figure 2.77 Plane view of the BJT for inverter-III (1000 V/1200 V 75 A chip, 16.5 mm × 16.5 mm).

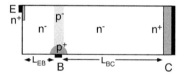

Figure 2.78 Basic structure of the BJT.

- The emitter n^+-region is shallow.[210]
- The width of the emitter n^+-region is a small part of the BJT's cross section.
- The base region consists of the p^--region and the p^+-region near the electrode.[211]

---

[210] It has the same tendency as the pin diode, indicating that the shallower p^+- or p-region causes good characteristics. Furthermore, it makes the analysis of the operating mechanism much easier.

[211] A grown-type BJT possesses such a p^+-region clearly. Furthermore, the p^+-region seems to be provided substantially because the impurities concentration of the p-region surface is high in any planar-type BJT.

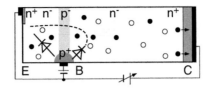

Figure 2.79 Basic operation of the npn-type BJT.

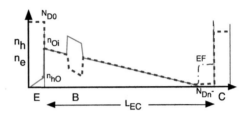

Figure 2.80 Charged carrier distributions of low-voltage/large-current operation (cross section in the upper side of Fig. 2.79).

- The n^--region exists between the emitter n^+-region and the base p-region.[212]
- A long n^--region exists between the base p-region and the collector n^+-region.

Figure 2.79 shows the operating situation of this BJT schematically. When the base is forward biased to the emitter, the densities of the holes and free electrons in the n^--regions on both sides of the base p-region increase because almost all regions except the collector n^+-region work as the pin diode, as shown in Fig. 2.41b.[213] In other words, the situation that there are a lot of holes and free electrons in the area sandwiched by the two n^+-regions appears.[214] If a positive voltage is applied to the collector with respect to the emitter, a free electron moves into the collector and a current begins to flow into the emitter through the n^--regions. That is, the BJT shows the resistance characteristic that is a straight line passing

---

[212] Such a model is permitted because the p-n diode can be regarded as the pin diode with the very short i-region, and its length does not affect the characteristics of the small current operation.

[213] In all n^--regions, $n_h \approx n_e \approx n_i \exp(qV_{BE}/2kT)$. Refer to Eq. 2.27, Section 2.4.3.

[214] Even in the p^--region the densities of the holes and free electron n_h and n_e also increase and those are expressed as $n_h n_e \approx n_i^2 \exp(qV_{BE}/kT)$.

Bipolar Junction Transistors | **127**

through the origin, $V_{CE} = 0$, $I_C = 0$.[215] This is exactly the core of the BJT operation.

By the way, there are two kinds of situations in the on-operation of the BJT. One is the situation of Fig. 2.80, that was explained as the core of the BJT operation in the last subsection, where a small positive voltage V_{CE} is applied to the collector of the npn transistor.[216] Its collector current I_{C1} is expressed in Eq. 2.61. D_e is the diffusion coefficient of the free electron in the n$^-$-region, L_{SB} is the width of the high-density area of the charged carriers,[217] and A_C is a cross section of the collector n$^-$-region. Besides, the charged carrier density on the emitter side edge of the n$^-$-region n_{Oi} is only determined by the base voltage V_{BE} independently of V_{CE} and is expressed as Eq. 2.62, which is substantially the same equation for the charged carrier densities in the i-region of a pin diode (Eq. 2.27).[218]

$$I_{C1} \approx 2q\, D_e \frac{\Delta n_e}{L_{SB}} A_C = 2q\, D_e \frac{n_{Oi}}{L_{SB}} A_C \tag{2.61}$$

$$n_{Oi} = n_i \exp\left(\frac{q\, V_{BE}}{2kT}\right) \tag{2.62}$$

In the n$^-$-regions where holes incline, an electric field EF_i, which exactly offsets the hole's diffusion motion, is generated. The coefficient 2 in Eq. 2.61 appears because the drift current of free electrons caused by this EF_i is just equal to their diffusion current.[219]

---

[215] I_C flows in the same direction even when $V_{CE} < 0$ as long as $V_{CE} > -V_{BE}$. Besides, for the electric field in the n$^-$-n$^+$ junction, the hole cannot move in the n$^-$-region.

[216] EF and N_{Dn-} indicate the EF and the donor impurity concentration in the n$^-$-region, respectively. The base is biased in the forward direction to the emitter.

[217] Its value changes according to the balance between V_{CE} and the collector current density J_C.

[218] Refer to Fig. 2.40.

[219] The balance of the diffusion current and drift current of holes is expressed in Eq. 2.63 as the drift velocity of the charged carrier and is $v_d = \mu\, EF$; μ means drift mobility. Then, the drift current of the free electron caused by EF_i becomes equal to its diffusion current as expressed in Eq. 2.64 using Einstein's relation, $q\, D = kT\,\mu$.

$$-q\, D_h \frac{dn_h}{dx} = q n_h \mu_h E\, F_i \tag{2.63}$$

$$-q n_e \mu_e E\, F_i = q n_e \mu_e \left(\frac{D_h}{n_h \mu_h} \frac{dn_h}{dx}\right) \approx q\, D_e \frac{dn_e}{dx} \tag{2.64}$$

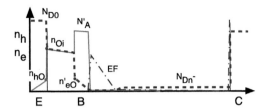

Figure 2.81 Charged carrier distributions of high-voltage/low-current operation (cross section at upper-side of Fig. 2.79).

In the other case, when a positive voltage V_{CE} of more than a few volts to a few dozen volts is applied, holes and free electrons on the collector side the n^--region are strongly pulled to the base and the collector, respectively. Then, the area appears between the base p-region and the collector n^--region where a fairly wide electric field exists and charge carriers are depleted, that is the so-called depletion layer, as shown in Fig. 2.81.[220] Free electrons in the p^--region move at a speed in proportion to the ratio of the density difference n_{e0}' to the width of the p^--region t_{p^-}. This is the collector current I_{C2} and is expressed as Eq. 2.65, where A_{p^-} is the cross section of the p^--region. As the section between the emitter n^+-region and the base p^--region operates as the pin diode, n_{e0}' is expressed as Eq. 2.66, where N_A' is the impurity concentration of the p^--region.

$$I_{C2} = J_C A_{p^-} \approx q D_e \frac{n_{e0}'}{t_{p^-}} A_{p^-} \qquad (2.65)$$

$$n_{e0}' = \frac{n_i^2}{N_A'} \exp\left(\frac{q V_{BE}}{kT}\right) \qquad (2.66)$$

In both cases the operating current of the pin diode between the emitter and the base is approximated to be $I_B \approx q D_h (\Delta n_h / t_n) A_n$ (Eq. 2.31), where A_n means the emitter electrode width, as explained at the end of Section 2.4.3. And this I_B value can be considered to be the base current in both BJT operations.[221]

---

[220] The inside of the pin diode between the base emitter is not affected by the collector voltage V_{CE}. The charged carrier density n_{oi} on the emitter side edge of the n^--region is the same as that before V_{CE} was applied.

[221] I_B is expressed as Eq. 2.67 concretely. Moreover, even for the BJT in a large current operation, it can be recognized that its pin diode between the emitter and the base performs the small-current operation. It will be easy to understand if thinking

Therefore, the current amplification factor, $h_{FE1} \equiv I_{C1}/I_B$, in a low-voltage operation for the first case is expressed as Eq. 2.68, which is equal to $L_{SB} \approx L_{EC}$. And the current amplification factor h_{FE2} of a high-voltage operation corresponding to the second case is expressed as Eq. 2.69.

$$I_B \approx q D_h \left(\frac{\Delta n_h}{t_n} \right) \quad A_n = q D_h \frac{A_n}{t_n} \frac{n_i^2}{N_{D0}} \exp \left(\frac{q V_{BE}}{kT} \right) \quad (2.67)$$

$$h_{FE1} \equiv \frac{I_{C1}}{I_B} \approx \frac{2 D_e}{D_h} \frac{N_{D0}}{n_i} \frac{t_n}{L_{EC}} \exp \left(\frac{-q V_{BE}}{2kT} \right) \frac{A_C}{A_n} \quad (2.68)$$

$$h_{FE2} \equiv \frac{I_{C2}}{I_B} \approx \frac{D_e}{D_h} \frac{N_{D0}}{N_A'} \frac{t_n}{t_{p-}} \frac{A_{p-}}{A_n} \quad (2.69)$$

Well, h_{FE2} is a constant because Eq. 2.69 contains only the numerical values that are fixed according to the transistor structure. This h_{FE2} is h_{FE} of the low-voltage transistor, illustrated in Fig. 2.61, right. Besides, in the case of low-voltage operation as shown in Fig. 2.80, the relation $h_{FE1} \propto 1/I_C$ is established because $I_C \propto \exp(q V_{BE}/2kT)$ is reduced from Eq. 2.61 and Eq. 2.62. It is well known that h_{FE} of a BJT becomes small in the form of $h_{FE} \propto 1/I_C$ in a large-current operation.[222]

By the way, the two operation modes indicated by Figs. 2.80 and 2.81 correspond whether the n^--region of the collector side has accumulated a lot of holes and free electrons or not. However, the tendency to fill up that area with holes and free electrons is also caused by increasing the peak charged carrier density n_{0i} of the n^--region even if V_{CE} is not small. Therefore, it is appropriate that Fig. 2.80 corresponds to the large-current, low-voltage operation and Fig. 2.81 corresponds to the small-current, high-voltage operation.

2.7.3 High-Voltage Large-Current Operation of the BJT

In the large-current, low-voltage operation as in Fig. 2.80, the n^--region near the collector n^+-region works as a resistor where the

---

about the behavior of the hole. By the way, n_{e0}' is determined basically by V_{BE} and does not depend on J_C.

[222] Furthermore, Eqs. 2.68 and 2.69 reveal that h_{FE1} and h_{FE2} are inversely proportional to the emitter cross section A_n. This characteristic is strange at a glance but is really observed.

130 | *Basic Technologies of Major Power Devices*

free electron density is N_{Dn-} and the collector current J_C, expressed as Eq. 2.70, is flowing.[223]

$$J_C = q N_{Dn-} v_d = q N_{Dn-} \mu_e E F \tag{2.70}$$

The increase of J_C is covered by getting the EF larger at first. However, the EF does not contribute to an increase of J_C when the EF gets large and the drift velocity v_d of the free electron gets closer to its saturated velocity v_s.[224] J_C gets merely proportional to the free electron density n_e (Eq. 2.71).

$$J_C = q n_e v_{s.e} \tag{2.71}$$

When J_C gets larger than $q N_{Dn-} v_{s.e}$, free electrons more than $v_{s.e} N_{Dn-}$ begin to flow into the n^--region close to the collector n^+-region. Then the negative charges exceed by $n_e - N_{Dn-}$ in that area. Because the EF obeys Poisson equation (Eq. 2.72), the EF increases toward the collector side with an incline in proportion to the net charges, $-n_e + N_{Dn-}$[225] (Eq. 2.73).

$$\epsilon_r \epsilon_0 \nabla^2 \psi = -q(n_h - n_e + N_D - N_A) \tag{2.72}$$

$$EF(x) \equiv -\frac{d\psi}{dx} + C = \frac{q}{\epsilon_s}(n_e - N_{Dn-})x + C \tag{2.73}$$

ϵ_r : Relative dielectric constant (12 : Si, 16 : Ge)

ϵ_0 : Dielectric constant in a vacuum (8.9×10^{-12} F/m)

C : Integration constant

---

[223] Free electrons of v_d N_{Dn-} flow into there from the base side. The EF is determined by adjusting the base-spreading region length, $x_b = L_{BC} - x_n$, to satisfy Eqs. 2.74 and 2.75. The first term of the right side in Eq. 2.74 is the inner voltage difference between the p^+-area and the n^--region. x_b and x_n are determined to satisfy Eqs. 2.76 and 2.77. Refer to Fig. 2.82 for x_b and x_n.

$$V_{CE} - V_{BE} = \frac{kT}{q} \ln\left(\frac{n_{p+}}{N_{Dn-}}\right) + E F x_n \tag{2.74}$$

$$V_{CE} \approx E F x_n \tag{2.75}$$

$$J_C \approx q \mu_e N_{Dn-} \frac{V_{CE}}{x_n} \tag{2.76}$$

$$J_C = 2q D_e \frac{n_p}{x_b} \tag{2.77}$$

[224] Figure 2.83 shows the electric field dependency of drift velocities and impact ionization factors.

[225] The current density that causes the distribution of the EF to be level is called J_0.

$$J_0 \equiv q N_{Dn-} v_{s.e} \tag{2.78}$$

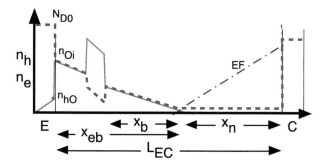

Figure 2.82 Charged carrier distribution in a short-circuit operation ($J_c > J_0$).

In the short-circuit test of the BJT, V_{CE} that is near the rating voltage is applied and J_C about three times larger than the rating current flows. Because the free electron density n_e flowing into the collector gets much larger than N_{Dn^-}, Eq. 2.73 is able to be approximated to Eq. 2.79.

$$EF(x) \approx \frac{q}{\epsilon_r \epsilon_0} n_e x = \frac{q}{\epsilon_r \epsilon_0} \frac{J_C}{q v_{s.e}} x \qquad (2.79)$$

Figure 2.82 shows distributions of the charge carrier densities n_h, n_e, and the EF at this time. Now, because the collector voltage V_{CE} is the area of the EF in the n$^-$-region, V_{CE} can be expressed as Eq. 2.81 using the maximum EF_m (Eq. 2.80).

$$EF_m \approx \frac{q}{\epsilon_r \epsilon_0} n_e x_n \qquad (2.80)$$

$$V_{CE} = \frac{1}{2} EF_m x_n \approx \frac{\epsilon_r \epsilon_0}{2q} \frac{EF_m^2}{n_e} \qquad (2.81)$$

Multiplying this equation and $J_C \approx q n_e v_{s.e}$ by each side leads to Eq. 2.82. This equation indicates that the value of $J_C V_{CE}$ at the short-circuit operation becomes a constant multiplication of a square of the maximum electric field EF_m (Eq. 2.82).

$$J_C V_{CE} \approx q v_{s.e} n_e \frac{\epsilon_r \epsilon_0}{2q} \frac{EF_m^2}{n_e} = \frac{\epsilon_r \epsilon_0 v_{s.e}}{2} EF_m^2 \qquad (2.82)$$

By the way, although the author finally noticed the basic operation mechanism of the BJT, written in the preceding section,

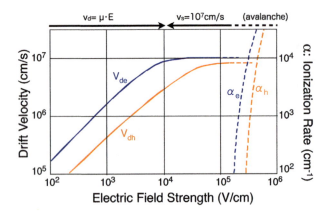

Figure 2.83 Electric field dependency of drift velocities v_{de} and v_{dh} and impact ionization factors α_e and α_h (measured values of Si at 300 K).

in about 2010 [21], the Eq. 2.82 was found and used to explain the spontaneous destruction mechanism in the short-circuit operation adopting the model in Fig. 2.82 in 1983 [3], that is, the free electrons that compose J_C cause collision ionizations and generate the pairs of holes and free electrons at the neighborhood of the maximum electric field EF_m portion. Those holes move into the base and work as a substantial base current. When the generation rate of the holes exceeds $1/h_{FE}$, I_C increases explosively because the increase of I_C by the current amplification action of the BJT and the generation of the holes that was proportional to n_e, that is $\propto I_C$, cause a positive feedback loop.[226] This is the mechanism of spontaneous destruction at the short-circuit test that always occurs at a constant electric power $J_C V_{CE}$, as shown in Fig. 2.71.

Collision ionization of the free electrons gets remarkable in an electric field higher than about 1.5×10^7 V/m, as shown in Fig. 2.83. And the generation rate of the holes becomes close to J_C/h_{FE} just at that EF. Therefore, if the current increases explosively from this

---

[226] Just before this positive feedback occurs, the stabilization mechanism works in a way that the spread base length x_b expands according to the increase of J_C and the effective h_{FE} decreases.

Bipolar Junction Transistors | 133

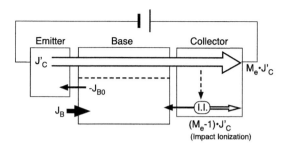

Figure 2.84 Basic principle of a BJT's high-voltage operation.

point, we get $J_C V_{CE} \approx 120 \text{ kW/cm}^2$ from $\epsilon_s \equiv \epsilon_r \epsilon_0 \approx 12 \times 8.85 \times 10^{-12}$ F/m and $v_{s.e} \approx 10^5$ m/s.[227]

$$J_C V_{CE} \approx \frac{\epsilon_s v_{s.e}}{2} EF_m^2$$

$$\approx \frac{106 \times 10^{-12} \cdot 10^5}{2}(1.5 \times 10^7)^2$$

$$\approx 120 \text{ kW/cm}^2 \qquad (2.83)$$

The high-voltage operation of the BJT can be indicated schematically, as in Fig. 2.84, paying attention to the charge carriers coming into and going out of the base region. M_e means the multiplication rate of a collector current in the n⁻ collector region as expressed in Eq. 2.84, where J_C' is the current coming in and J_C is the final current. And $J_{B.net}$ is defined as the net hole current flowing into the base region (Eq. 2.85), which consists of three factors, J_B, J_{B0},[228] and $(M_e - 1)J_C'$, which correspond to the current flowing in from the base electrode, flowing out from the base region, and flowing in from the collector side, respectively, (Eq. 2.85).

$$M_e \equiv \frac{J_C}{J_C'} \qquad (2.84)$$

$$J_{B.net} \equiv J_B - J_{B0} + (M_e - 1)J_C'$$
$$= J_B - J_{B0} + \left(1 - \frac{1}{M_e}\right)J_C \qquad (2.85)$$

---

[227] J_C of Fig. 2.71 was induced using the emitter area of the last stage of a Darlington transistor. If using its chip area, the critical value corresponds to $J_C V_{CE} \approx 100$ kW/cm².

[228] $J_{B0} = J_C/h_{FE}$ for a low-voltage operation.

134 | *Basic Technologies of Major Power Devices*

Table 2.2 Stability of a BJT's high-voltage operation

$\Delta M_e/\Delta J_C$	$J_{B.net}$	Judgement	#
(M_e uncertain)	—	destruction-A	I
positive	positive	destruction-B	II
	negative	(destruction-C)	III
	negative	stable operation	IV
negative	positive	stable increase	V
	negative	stable decrease	VI

And, it was thought that the BJT led to destruction necessarily in a situation that satisfied the next two conditions at the same time because the collector current of the BJT must continue to increase.

- The ratio of a change in the multiplication factor M_e to that in the collector current is positive, namely $\Delta M_e/\Delta J_C > 0$.
- The inflow of the hole into the base region is superior to the outflow from it, namely the net base current $J_{B.net} > 0$.

Then, depending on the value of $\Delta M_e/\Delta J_C$ and $J_{B.net}$, the stability of a BJT in the high-voltage and large-current operation could be judged as in Table 2.2 [20]. Basically, a stable operation would exist even under a high voltage if $\Delta M_e/\Delta J_C < 0$ was established. And $\Delta M_e/\Delta J_C < 0$ is always valid if J_C is less than $J_0 \equiv q(N_{Dn^-})v_{s.e}$. It is because the collision ionization rate becomes small due to a decrease in the maximum electric field EF_m, which is caused by an increased J_C, namely increased negative charges, to lessen the slope of the EF as shown in Fig. 2.85.[229] Then, the hole generation rate by collision ionization in the n^--region gets small and M_e decreases, too.[230] This really corresponds to the explanation of Fig. 2.72, namely that the BJT did the sustain operation and was not destroyed if J_C was less than J_0. The sustain operation is exactly the state of $J_{B.net} = 0$ and $\Delta M_e/\Delta J_C < 0$.[231]

---

[229]The electric field distribution EF(x) is induced as in Eq. 2.73 to satisfy the Poisson equation (Eq. 2.72).

[230]If $J_C > J_0$, the situation of Fig. 2.86 appears and it is established that $\Delta M_e/\Delta J_C > 0$.

[231]The operating points, namely trace, indicated in the area of $I_C \leq 10$ A in Fig. 2.72 are called the sustain waveform, where the BJT sustains a certain voltage that corresponds to the flowing current value by itself. These operating points are

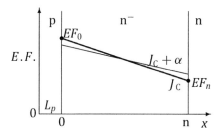

Figure 2.85 Effect of ΔJ_C on the EF in the n$^-$-region–I: at $J_C < J_0$.

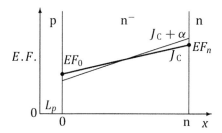

Figure 2.86 Effect of ΔJ_C on the EF in the n$^-$-region–II: at $J_C > J_0$.

In addition, that a voltage range expands with an increase in the reverse base current, as shown Fig. 2.72, against conventional common sense would be understood from the viewpoint of the second point above.

On the other hand, if a calculation for M_e diverged,[232] it could be regarded as a destruction by considering that a stable operating condition did not exist. The spontaneous destruction in the short-circuit operation mentioned before would be caused by this mechanism.

2.7.4 Safe Operating Area of the BJT

The SOA of a BJT is expressed as shown in Fig. 2.72. In its operation where $J_C > J_0 = qN_{Dn^-}v_s$, which is indicated as a horizontal dashed

---

the border of rows V and VI in Table 2.2. The calculation lines in Fig. 2.72 were calculated to the combinations of J_C and V_{CE} using the author's own program.

[232] It corresponds to the row I in Table 2.2.

line, a BJT must be destroyed inevitably if the operating point (J_C and V_{CE}) became the upper-right side of the limit line inclined right downward.[233] In the case of $J_C < J_0$, the BJT would be destroyed instantly when the V_{CE} is more than the upper limit voltage of the sustain waveform, which depends on the reverse base current I_{B2}.[234]

The second breakdown phenomenon of the BJT has been a problem for a long time but it is caused by a parasitic structure part in the chip. If this parasitic part is removed from the BJT, a simple principle functions in it, that is the operating limit is determined by a polarity of $\Delta M_e / \Delta J_C$ in the collector n^--region and a balance between incoming and outgoing holes in the base region. Its simplicity is caused by (a) the BJT has basically a 1D structure and (b) only one kind of charged carrier runs in the high-electric-field region.

Furthermore, the BJT possesses the mechanism to restrain the current increase until the moment when positive feedback for the collector current explosion begins to enter. Therefore, the BJT without a parasitic structure is able to realize the operating limit close to the theoretical value even if the chip area gets large.

The IGBT has a much larger operating area than the BJT, but the mechanism to determine its operating limit is much more complicated because it does not satisfy the (a) and (b) factors mentioned above.

2.8 Metal-Oxide-Semiconductor Field-Effect Transistors

The lateral MOSFET was developed in the beginning of the 1960s and is a semiconductor device used the most as a basic unit of the ICs now. Nevertheless, it was regarded not to be suitable for the power device because its operating capability was poor. However,

---

[233] This corresponds to the spontaneous destruction in the short-circuit operation.

[234] First, an operating point of the BJT moves upward along its sustain waveform by the counter electromotive force. If that force goes down in the way of that process, the operating point shifts down on the sustain waveform and the BJT turns off safely. When it was not so, the operating point reaches the upper limit of the sustain waveform and the BJT would be destroyed.

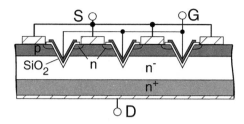

Figure 2.87 Schematic diagram of a V-groove MOSFET.

in 1975, the 60 V 2 A product that has many MOS gates on exactly the identical V-grooves[235] in a high density, as shown in Fig. 2.87, appeared and the vertical MOSFET attracted a lot of attention as a power device. Using this technique, precise, and expensive equipment was unnecessary for production because the gate length whose control was the key of the MOSFET production was fixed by the difference between the depths of the p-region and n-region. In addition, there was the advantage that the effective area of the chip increased because a drain electrode was established on the whole undersurface of the chip. Because there were some problems, for example, the electric field was concentrated on the tip of the V-groove and the etching process was not stable, it was not able to become a mainstream device. However, it was significant for the vertical MOSFET to have a possibility as the power device anyway.

The diffusion self-alignment metal-oxide-semiconductor field-effect transistor (DSA-MOSFET) shown in Fig. 2.88 was developed in the field of IC two years after the V-groove MOSFET. It forms the surface portion using the depth difference of the twofold diffusion of p-type and n-type impurities through the same hole of a SiO_2/polysilicon layer, which works as a gate oxide and a gate electrode ultimately. These techniques that can enable one to make a minute gate structure easily were diverted to the power MOSFET as it is and spread rapidly from the late 1980s with the progress of miniaturization.

The MOSFET that uses this diffusion self-alignment technology is called the DSA-MOSFT or the D-MOSFET. And most MOS devices

---

[235]Those were formed using silicon's characteristic that a (111) face can be exposed exactly by etching in a potassium hydroxide solution.

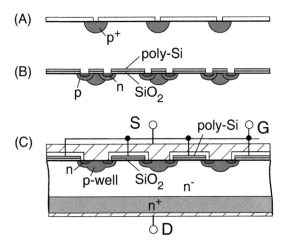

Figure 2.88 Vertical DSA-MOSFET and its manufacturing process.

except the trench-MOSFET/IGBT, which will be explained later, have used this technology.[236] With this structure, the gate peripheral length must be enlarged to get a low on-voltage first. For this purpose, each cell pattern is designed to be quadratic or hexagonal and the shallow p-well is very effective because it enables the placement of the cells in a high density. Making the p-well shallow brings about the effect of shortening the gate length, too. Besides, the polycrystalline silicon that is functioning as a gate electrode surrounds the all p-well and is connected to a gate lead electrode around a chip.

The trench-MOSFET shown in Fig. 2.89 has been widely used in the field of low withstanding voltages, of less than several dozens of volts, from the late 1990s in particular. The trench, namely a U-groove, is dug perpendicular to a silicon wafer surface with a reactive ion etching that uses the plasma gases, and the MOS gate is formed on the side of this groove. Although the planar pattern of the V-groove MOSFET using the anisotropic etching of the silicon wafer is limited to a linear pattern, any trench-gate pattern can be formed

---

[236] Planar-IGBTs use this technology too.

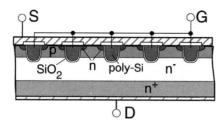

Figure 2.89 Schematic diagram of a trench-gate vertical MOSFET.

according to its mask pattern.[237] Besides, the groove pitch can be produced extremely narrow (less than 1 μm). Therefore, the MOS gate density can be increased to a value much higher than that of the planar-gate structure. For the low-voltage MOSFET, the trench-gate structure is advantageous because the high-resistivity n^--region is designed as thin as possible and the contribution of the MOS gate part in the on-voltage is large.

By the way, the demand for a DC-DC converter working with a low-voltage large current is rapidly growing because the large power consumption of a fine-dimensional IC represented by the central processing unit (CPU) of a PC came to need a distributed power supply recently. For this application, a low-voltage MOSFET (less than about 20 V) is used, and there is movement to use the planar-type MOSFET, which was considered to be unsuitable for the power application. It is because the gate peripheral length density can be larger than the trench-gate structure by the progress of the miniaturization technology of the plane pattern.

Figure 2.90 shows the schematic cell structure of an advanced planar-MOSFET as the typical example. The n^- offset-region extends from the drain n^+-region, and a gate electrode is formed over there. This n^--region is set so that the trade-off between a withstanding voltage and an on-voltage becomes more suitable. This structure is used for a high-frequency MOSFET because of the small gate capacity. The planar device is also suitable for a high-density assembly as the electrode is concentrated on one side. To reduce the power loss and stray inductance by the wiring, a chip with multilayer

---

[237] The trenches are formed in a mesh pattern and the whole polysilicon in those is connected to a gate drawer electrode around a chip.

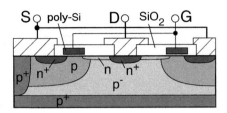

Figure 2.90 Tiny lateral MOSFET (offset gate structure).

electrodes is sometimes mounted directly on a printed board via a lot of ball-formed metals.

Although it is natural for there to be a demand for a lower on-voltage in a MOSFET as a power device, there are many uses in which high-speed operation is needed in the MOSFET as it is the highest-speed switching device, for example, for the switching power supply, the final stage amplifier of cell-phone transmitter, and the power supply of the pulse driven laser.[238]

The MOSFET could be regarded to consist of a variable resistance and capacitors. The current flowing in a resistance is equal to the current of the channel part and is determined entirely by the gate potential V_{GS}. As the displacement current flowing through the capacitor can be ignored compared to the main current, the operating current of the MOSFET is determined uniquely by the gate potential V_{GS} for each moment. Because the gate is isolated electrically, it can be treated as a capacitor. Charging and discharging speeds determine the on-time and off-time of the MOSFET. And the trench-gate structure is unsuitable for a high-speed operation as its gate capacitor is large.

In any device, the MOSFET has reached the theoretical limits of its estimated characteristics, achieved using the same manufacturing method as an ultra-large-scale integrated circuit (ULSI). However, the superjunction (SJ) MOSFET, with characteristics beyond these limits was manufactured in 1998. Figure 2.91 shows this manufacturing method. Before the usual DSA-MOSFET process

---

[238] In one instance, dozens of MOSFETs had to be made connected in a series and in parallel in the laser power supply, which needed dozens of kilovolts. Its on-voltage was more than dozens of volts, but a steep pulse wave pattern could not be provided with other devices.

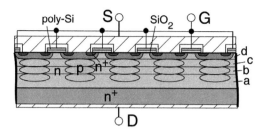

Figure 2.91 Example of a superjunction MOSFET structure.

on the wafer surface, epitaxial growth layers, whose impurity concentration was orders of magnitude higher than that of the usual MOSFET, had been formed divided into several times, and in each layer the p-region was formed by the ion implantation method.[239] As a result, vertically long n-region and p-region are laminated laterally in the drain region.

If the impurity amount of the p-region is approximately equivalent to that of the n-region and if both regions are sufficiently thin, whole areas are depleted when the reverse voltage is applied and behave as a nearly intrinsic area because the negative ions of the p-region and the positive ions of the n-region cancel out each other. Then, the whole drain region becomes a substantial depletion layer[240] and can hold a high withstanding voltage. This situation can be valid theoretically even if the impurities concentration of the n-region is several orders higher than that in a normal MOSFET. Because the on-current drifts through this n-region, the on-voltage can be largely improved. This structure is effective in improving the on-voltage of a several-hundred-volt MOSFET by nearly one digit. The idea of this structure has existed for several decades, but improving the precision of the device and the process simulations must be necessary to realize it as a product.

The destruction of a MOSFET is divided into the destruction of the MOSFET by itself and the destruction caused by the built-in pin diode[241] operation in the vertical MOSFET.

---

[239] Furthermore, additional ion implantation of the n-type impurity might be done between the p-regions.

[240] That is, the whole drain region can be recognized as an i-region.

[241] Refer to Figs. 2.88, 2.89, and 2.91.

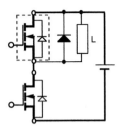

Figure 2.92 Chopper circuit using a vertical MOSFET.

It may be said that the second breakdown does not happen because the MOSFET is a unipolar device. Certainly a MOSFET is stable thermally and is not easy to be destroyed because of its poor conduction ability compared to a BJT and an IGBT. However, the MOSFET might lead to the thermal runaway if it operates while being connected to a constant current source. In that case, it is because the positive feedback phenomenon of the temperature rise might occur as the on-resistance of the MOSFET increases with temperature. Moreover, the second breakdown is possible potentially because any MOSFET definitely contains a parasitic BJT portion.

The built-in pin diode becomes a problem when the vertical MOSFET is used in a chopper circuit or an inverter circuit as shown in Fig. 2.92. When the lower-arm device is turned on after a circulating current begins flowing through the built-in pin diode, the recovery operation of this diode is much slower than the original FWD because a high-speed diode is ordinarily used as the FWD. If the rising rate of the drain-source voltage V_{DS} is steep, the power loss caused by this delayed current is not negligible. In addition, the residual charges that were not exhausted in the recovery operation start the parasitism n^+p-n$^-$ transistor consisting of a source region, a p-well, and an n$^-$-region and it is possible to involve these in the second breakdown. To avoid this problem, a high-voltage Schottky barrier diode (SBD) made from SiC might be used for a FWD or a low-voltage SBD might be inserted to prevent the circulating current from flowing into the MOSFET, as shown in Fig. 2.93.

However, in recent years an inverter circuit omitting the FWD was usual in a low-voltage use for cost reduction and such a

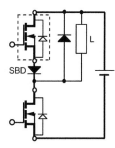

Figure 2.93 Chopper circuit using a vertical MOSFET and a SBD.

tendency began to expand to the 300–400 V class MOSFET. For that purpose, a MOSFET in whose case it is difficult for the parasitic n^+pn^- transistor operation to occur must be chosen. In other words, the MOSFET must be one in which the width of the source n^+-region is narrow and the impurities concentration of the p-well under the source region is high.

In addition, during the off-operation in unclamped inductive switching, which drives the L-load by MOSFET without a clamping circuit,[242] the MOSFET needs the ability to hold directly the high voltage that the L-load generates while allowing almost the same current to flow as in an on-operation. This destruction is the avalanche destruction explained at the end of Section 2.3.3. Because the decrease of the current flowing L-load, which produced the reverse voltage, is caused by closing the channel, the MOSFET does not work already. Therefore, the avalanche operation occurs in the pin diode inside the MOSFET. Although a pin diode has a particular avalanche withstanding ability, as explained in Section 2.3.3, the MOSFET chip would be destroyed below its limiting current. It is because the parasitic n^+pn^- transistor that the MOSFET has built into the source side is turned on by the avalanche current. So it is an important requirement of the power MOSFET to suppress an operation of the parasitic n^+pn^- transistor.

The other important problem of a MOSFET is the reliability of the gate oxide. It can be classified into long-term reliability and

---

[242] The circuit that prevents the voltage rise above a certain value not to be added to the switching device.

surge-withstanding ability. The former is a problem similar to the insulation deterioration of a high-voltage cable. A slight leak current flows when a high electric field is added to the gate oxide; then it deteriorates with an increase of integrated charges. Of course, it is produced with enough margin, but the gate leakage current cannot be screened severely at the p-n junction because a large current might leave permanent damage. Nevertheless, the surge-withstanding ability is not much of a problem with the power MOSFET.[243] If presuming a measure, a protection diode that is sometimes made by polycrystalline silicon on a chip is attached between its gate and source.

For the MOSFET, various abbreviated designations are used as follows: L-MOSFET (lateral MOSFET), V-MOSFET (vertical MOSFET), V-groove MOSFET, U-MOSFET (U-groove MOSFET = trench-MOSFET), D-MOSFET (DSA-MOSFET), and LD-MOSFET (lateral D-MOSFET).

2.9 Insulated Gate Bipolar Transistors

2.9.1 IGBT Structures

It is a generally accepted opinion that an IGBT, which is the most successful power device, has the structure to drive a pnp transistor by a MOSFET and it was invented in the United States in about 1980. However, the device of such a constitution as is shown in Fig. 2.94 was applied for in Japan in 1968 and it became a patent in 1972 [22].

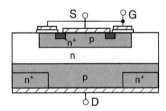

Figure 2.94 MOSFET-driven pnp transistor structure (Yamagami and Akagiri's patent).

---

[243] A protective circuit is incorporated in an input or output terminal as a measure with the IC.

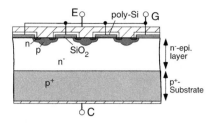

Figure 2.95 Basic structure of an IGBT (NPT IGBT-I).

Various development groups gave it various names, such as COM-FET (conductivity modulated field-effect transistor), IGR (insulated gate rectifier), IGT (insulated gate transistor), and BiFET (bipolar mode MOSFET), before it was named IGBT in 1986. IGR likened it to diode operation [52], but others announced a principle to drive the pnp transistor by a MOSFET.

Figure 2.95 shows a schematic diagram of the first IGBT produced experimentally. It was made from a wafer in which the n^--epitaxial layer was grown on a highly-doped p^+-substrate. A device with the n-buffer, as in Fig. 2.96, and the IGBT, which used a lifetime control process, was also produced experimentally before long. It was by adopting the n-buffer that the thickness of the n^--region could be reduced drastically and resulted in a much lower on-voltage, keeping the withstanding voltage same. The lifetime control was introduced to shorten the long off-time, 10–50 µs, of the early IGBTs.

The structure that established the n-buffer has been called the punch-through (PT) structure as the depletion layer of the electric charge can be expanded from the emitter side to the n-buffer when holding a high voltage between the collector and the

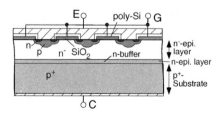

Figure 2.96 Basic structure of an IGBT (PT IGBT).

emitter. In contrast, with the structure without the n-buffer, the n^--region must be designed to be thick enough not to cause the leak current to increase remarkably when the depletion layer approaches the collector p-region.[244] This non-punch-through (NPT) structure appeared again in about 1990, as mentioned later, together with the new technology that does not use an epitaxial growth wafer.

From the first, it was well known that a parasitic pnpn portion works as a thyristor, namely "latch-up," and this was recognized as the greatest problem of the IGBT.[245] The developing story of the IGBT was the history to suppress this thyristor action while improving the trade-off between the on-voltage and the switching properties. The lifetime control was effective as a measure to suppress this action. The 1000 V IGBT module was manufactured in 1986 by Toshiba[246] and it was used in an AC 480 V 15 kHz switching inverter. It contained the lifetime-controlled IGBT chip whose emitter-cell pattern was changed from a polygon to a stripe to reduce the gate peripheral length and the impurities concentration of the p-well was increased to suppress the latch-up.

However, it did not spread much because the on-voltage was larger than that in the BJT. It was in the early 1990s that the IGBT module of the 600 V class surpassed the transistor (BJT) module of the Darlington connection. The IGBT, which should have a basic structure more suitable for a high voltage than the BJT, had conflicted with the BJT, besides the Darlington connection, at the most severe 600 V class in cost because the IGBT could not reduce its chip size as the current-carrying ability rapidly decreased with an increasing withstanding voltage.[247]

The on-voltage has been improved in every generation to make the p-well shallower, as shown in Fig. 2.97. For that purpose, miniaturization of the emitter pattern was necessary, but miniaturization in itself did not contribute to the reduction of the on-voltage unlike the MOSFET. Although the IGBT works as "the partial pin diode that has a current control structure in the main current

---

[244] Refer to Fig. 2.24.

[245] The n^+-regions on the collector side in Fig. 2.94 of Yamagami and Akagiri's patent was aimed to prevent it.

[246] In 1985, a 600 V, 25 A module that contained a single IGBT chip was released.

[247] It is in recent years that its reason was explained clearly [23].

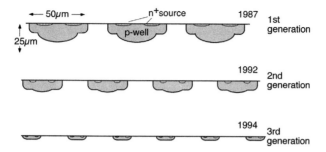

Figure 2.97 Improving the IGBT's on-voltage by reducing the p-region ratio.

path," the partiality increases if the p-region of the emitter side gets larger, especially deeper [23], as explained later, in Section 2.9.2.

Improvement of the IGBT is carried out energetically even now, more than 30 years after its invention, whereas technology development of the BJT for the inverter had matured over several years. In past development, the manufacturing technology of the wafer (see the list below) and the design technology of devices (see the list after the four points below) were important.

Manufacturing technology on wafer

- Increasing precision and reducing price of the epitaxial growth technology, which enabled the n-buffer structure
- EB irradiation having been effective and handy as a lifetime control method
- A trench-gate structure that advanced a low on-voltage (trench-IGBT) [53]
- Evolution of the wafer process technology to enable the use of an extremely thin wafer, 50–100 μm, [54]

Design technology on devices

- The protection method of a thyristor action, namely latch-up (high impurity doping in the p-well, stripe-emitter pattern, and lifetime control).
- A shallow emitter side p-region that keeps the on-voltage remarkably low.

- An NPT structure without lifetime control, which was started in more than 1000 V, which used a single crystal wafer [25].
- New collector side design considering the hole and free electron distributions inside an IGBT, which realized an IGBT of more than 3000 V [25].
  The combination of a low-dope n-buffer and shallow low-dope p-collector region enabled shortening of the n^--region length [26, 27].
- New collector side design considering the hole and free electron distributions inside an IGBT—injection-enhanced insulated gate transistor (IEGT) [28], carrier-stored trench-gate bipolar transistor (CSTBT) [29], high-conductivity insulated gate transistor (HiGT) [30], and field stop insulated gate bipolar transistor (FS-IGBT) [31]).
- A reverse-blocking IGBT that has the withstanding voltage for the emitter to the collector and is aimed mainly for the matrix converter [32].
- Packaging technology on the low stray inductance, downsizing, and long-term reliability.

Table 2.3 shows these technical appearance times and combinations. Improvement has been carried out while winding based on the existing technologies of the pin diode and the GTO.

Figure 2.98 shows the NPT insulated gate bipolar transistor (NPT-IGBT) structure that overturned the common sense beliefs as per which the n^--epitaxial layer, the n^+-buffer, and lifetime control were necessary in the IGBT, until about 1990. It was like the first IGBT of Fig. 2.95 but it was different in that the high-resistivity n-type wafer without the expensive epitaxial layer was used and the

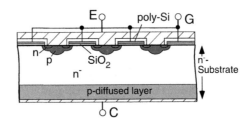

Figure 2.98 Basic structure of the IGBT (NPT IGBT-II).

Table 2.3 Development history of the IGBT structure.

	code	NPT-I	PT-I	PT-II	PT-III	NPT-II	H.V.	F.S.	IEGT	(IEGT-p)	CSTBT	HiGT	reverse blocking
substrate epitaxial crystal	A	○	○	○	○		○		:	○	:	:	○
single crystal	3					⊙		⊙	:		:	:	○
thin single crystal	D								:		:	:	○
life time control electron beam rad.	B		⊙				△		:	○	:	:	
local control							△		:	△	:	:	
collector structure n⁺-buffer			⊙	○	○				:	○	:	:	
light n⁺-buffer	4						⊙		:		:	:	
high dope p⁺-sub. p-diffusion		○	○	○	○	⊙		○		○			○
shallow p-diffus.	4					△			:		:	:	
emitter pattern polygon	1	○	○						○	○	○	○	○
stripe	2			⊙					○	○	○	○	○
emitter structure shallow p-well	C	○								○	○		
trench gate									○		○		
trench ratio large	5				⊙			:					
trench + n-region								:	○		⊙		
p-well + n-region									⊙			⊙	
reverse blocking	6												⊙

(⊙: new technology applied, ○: applied, △: applicable, ∷: sellectable)

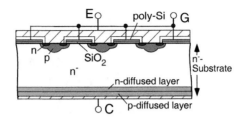

Figure 2.99 Basic structure of the IGBT (HV-IGBT, FS-IGBT).

collector p-region was lightly dosed by diffusion. Until the latter half of the 1990s, it was adapted exclusively to 1.2–2 kV class IGBTs using a wafer of about 200 μm thickness because a thin wafer of about 100 μm was too fragile to be treated safely in a wafer process. Because the n⁻-region was thicker than the mainstream PT insulated gate bipolar transistors (PT-IGBTs), their destruction-withstanding ability was large, but on the other hand the on-voltage was higher than those. But there was the advantage that it did not use the lifetime control process that caused a large temperature dependency of the electric characteristics.

Figure 2.99 shows the collector structure of the high-voltage insulated gate bipolar transistor (HV-IGBT), which has characteristics resembling those of the GTO of more than 3 kV. The key was to shorten the n⁻-region length to prevent increase of the on-voltage. This was accomplished by establishing a low-concentrated n-buffer and holding the withstanding voltage with the pin structure.[248] To get good turn-off properties, the supplying ability of holes from the collector side was reduced by lowering the impurity dose and thinning the length of the p-collector region. It had been found that such a collector p-region structure was effective for high-speed operation during the development of the NPT-IGBT [33], but it was not applied because the on-voltage must increase. And this was realized after thinning the n⁻-region by adopting a pin structure. Although lifetime control is not necessary, in a high-voltage IGBT local lifetime control is effective, particularly to improve the switching loss, because it can optimize the charged

---

[248] This was the well-known method for a long time for high-voltage diodes and GTOs. Refer to Section 2.3.1 for a comparison of a pin structure and a pip-structure.

carrier distributions in its long n^--region. This structure is called the soft-punch-through IGBT or light-punch-through IGBT.

The FS-IGBT adopted this structure to an IGBT less than about 1 kV using the technology to manufacture a thin wafer less than about 100 μm in the processing field of the IC. As the n^--region is short in a low-voltage IGBT, charged carrier distributions of the n^--region can become adequate and enough characteristics are provided if the supplying ability of holes from the collector side is regulated without lifetime control. It is attractive from the point of view that it is a structure that consists of the minimum elements necessary for the IGBT operation. However, enough attention must be paid for its poor destruction-withstanding ability to be reduced on a sudden temperature rise in the short-circuit operation, for instance. That is, thinning a wafer causes the IGBT to easily increase its temperature because its heat capacity becomes small and the solder between the IGBT chip and a heat sink behaves as an insulator for a very short time. So the chip temperature becomes very high at the short-circuit operation, and it causes the leak current to increase, and this process could continue recursively until the thermal runaway. It is a situation in which a current increases abruptly after dozens of microseconds during the off-state and leads to destruction, similar to the waveform-C in Fig. 2.122[249] [34].

By the way, Figs. 2.100 and 2.101 are schematic diagrams of a trench-IGBT. The trench is formed in line-segment shape along the depth direction unlike the case of a MOSFET. At first, the trench-IGBT was expected to become the ultimate IGBT because it could increase the channel width a lot, but the following IEGT had proved that the on-voltage got lower when the channel density was restricted to a certain value. Besides, the on-voltage also decreased even if attaching the n-region under the p-region between the trenches.[250] A trench-gate structure cannot be an effective measure by itself to improve the on-voltage and would be understood to be one of the elemental technologies of IGBT design.

---

[249]This short-circuit destruction mode rarely happens in the BJT but is easy to be generated in an IGBT that is made of thin wafer, such as FS-IGBT, which operates in a high current density.

[250]That was proved by the following CSTBT.

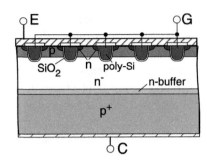

Figure 2.100 Schematic diagram of a trench-IGBT.

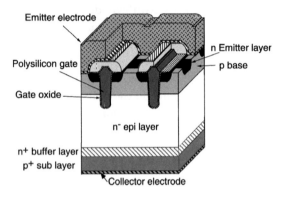

Figure 2.101 3D cross-section diagram.

IGBT has the characteristic that the charged carrier density on the emitter side in the n^--region always decreases. As this is a basic characteristic of the IGBT, its on-voltage becomes larger inevitably than that of a thyristor-type device.[251] The IEGT, CSTBT, and HiGT are the results of activities to overcome this disadvantage. Figure 2.102 shows the aimed effect schematically.

The IEGT drawn schematically in Fig. 2.103 was a pioneer of the improvement that paid attention to this point.[252] Each of following

---

[251] Its charge carrier density on the n^--region is basically uniform.

[252] The development of IEGTs had started when a new enhancement effect of conductivity was found in the research activity to drive a high-power GTO by a gate voltage instead of using a large gate-driving current. And that effect was found to be applicable also to the IGBT structure in 1993 [28].

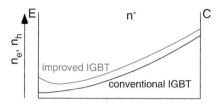

Figure 2.102 Charged carrier distributions of IGBTs in the on-state.

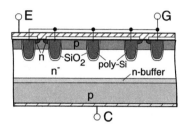

Figure 2.103 IEGT: Reducing trench-gate IGBT.

three factors was finally discovered to be effective in improvement of the charged carrier distribution and resulted in a low on-voltage.[253]

- Lengthening a cropping-out portion to the n^--region of the trench
- Establishing the source-region at intervals
- Narrowing the width of the mesa-part[254] where the source exists

The improvement points can be summarized as (i) reducing the placement density of the trench gate appropriately, and (ii) reducing the emitter side p-region area that is neighboring the n^--region or floating on it electrically. The appearance of the IEGT gave a shock by overturning the common sense thought, "The higher the trench-gate density the lower the on-voltage that can be achieved." Afterward, this IEGT structure was understood to be effective in all trench-IGBTs unrelated to the voltage class and it has been adopted for most of the latest 600 V–2 kV trench-IGBTs.

---

[253] In addition, the structure that made the width of the trench ditch wide was also fabricated.
[254] That is the "mesa" that is sandwiched in trenches.

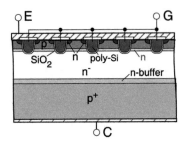

Figure 2.104 Schematic diagram of the CSTBT (emitter portion).

The trench gate was not adopted for a long time except in the IEGTs of Toshiba and in IGBTs of more than 3 kV. It has been said that the collector structure, particularly thickness, was the first factor in IGBTs of several kilovolts and the advantage of the trench-gate did not exceed the rise in cost, but the application finally began in late years [35].

In addition, someone called the several-thousand-volts planar-IGBT the "planar-IEGT," but this name is not appropriate because its fine on-voltage was caused simply by reducing the p-area on the emitter side, as shown in Fig. 2.97 and it had the same structure as a HV-IGBT, indicated in Fig. 2.99.

In the CSTBT is an established low-concentrated n-region outside of the p-region in a mesa area, which is between the trenches, as shown in Fig. 2.104. The on-voltage becomes lower so that the impurity dose of this n-region is higher. This came about with the intention to bring the IGBT operation closer to the pin diode as much as possible[255] [29]. It can be combined with thinning out source regions characterized by the IEGT.

The HiGT applies a principle of the CSTBT to a normal double-diffused metal-oxide-semiconductor (DMOS) emitter structure and establishes the low-concentrated n-region outside of its p-well region. It is easy to process, and then it is applied to some high-voltage IGBTs to decrease the on-voltage.

In addition, in recent years the structure has been used to establish the n-region only around a channel opening portion of the

---

[255] Of course, the withstanding voltage decreases if the doping is too much.

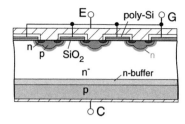

Figure 2.105 Schematic diagram of the HiGT (emitter portion).

p-well in several-kilovolt IGBTs [36]. This structure can reduce the tendency to degrade the withstanding voltage while improving the on-voltage enough.

In each generation in the IGBT development the following technical efforts have been made. The first problem was reduction of the on-voltage across each generation.

~**1990:** Optimization of the n^+-buffer and lifetime.
~**1995:** Refinement of the emitter pattern.
   In fact, it was an effect of reducing the p-region ratio on the surface.
~**2000:** Expansion of the channel width by using the trench gate.
   However, the channel width, namely the channel density, was not the first factor and there was the suitable p-well density.[256] The 3–4.5 kV IGBT has been in use by realizing the desirable charged carrier distribution in the n^--region (light-punch-through structure).
~**2005:** The most suitable design that utilized the thin wafer processing technique.
   Even in the low-voltage IGBT, the desirable charged carrier distribution in the n^--region has been realized.

It may be said that the aim of each generation was not necessarily right. The NPT-IGBT proved that n^+-buffer and lifetime control were not necessarily needed and refinement of the emitter pattern did not contribute by itself but contributed to the reduction of the p-region size, namely influence, on the emitter side. The trench-IGBT was

---

[256] Reduction of the emitter side p-region was effective, after all.

said to be the ultimate IGBT because it had the advantage that the channel width could be increased, but the on-voltage was increased by limiting the trench density (IEGT).

Various IGBTs are manufactured now. The key technologies to make them are summarized as follows.

- Stripe-shaped emitter pattern.
- Increasing the charged carrier densities on the emitter side.
  - Establishing the n-region outside of the p-region (CSTBT, HiGT).
  - Reducing the p-region on the emitter side in the planar-IGBT, namely making the p-well shallow.
  - Narrowing the mesa width and increasing the mesa height in the trench-IGBT.
  - With the trench MOS structure, thinning the gate portions and floating p-regions without the gate from the emitter electrode electrically (IEGT).
- Decreasing the supply of holes from the collector side.
  - Balancing the doses and thicknesses between the n-buffer and the p-region for the PT-IGBT.
  - Making the collector p-region shallow and in low dose for the NPT-IGBT.
- Shortening the collector n^--region (LPT-IGBT).

After all, it seems wise to conclude that a special IGBT would not overwhelm others and adjusting the charged carrier distributions adequately is important. For this purpose, it is important to bring the IGBT, which is the partial pin diode, close to the ideal pin diode as much as possible, particularly by reducing the influence of the p-region on the emitter side [23].

The withstanding voltage of the IGBT has been realized up to 6.5 kV 600 A in a module package. In the 4.5 kV class there are a 900 A module and a 2.1 kA flat package.[257] These devices are used as inverters for DC 3–4 kV lines, for instance. The on-voltage of the GCT is lower in this area, and there are still few uses of the IGBT. However,

---

[257]2.1 kA is not the root-mean-square (RMS) on-state current rating but the maximum controllable current.

the IGBT has replaced the GTO in the field of electric trains because the IGBT does not need an anodal reactor for short-circuit protection and its gate drive circuit is much simpler and easier. Furthermore, the IGBT has begun to be used even in large equipment, such as equipment for steel manufacture. In these field the GCT and the IGBT would coexist in the future.

However, there is no trend toward developing an IGBT of more than 6.5 kV. This is because it would be close to the silicon limit and there would be only small uses besides. Further improvement of the destruction endurance ability and the long-term reliability or raising the operating temperature limit more than 125°C would be the issue rather than increasing the operating voltage.

The IGBT would be said to be a device that switches on and off a partial pin diode directly using the MOSFET connected to it in a series. The partiality is because there is extra p-region in its emitter area, which should act originally as the n-type cathode of the pin diode [23].[258]

The IGBT has the following characteristics:

- High-speed operation with low driving power is possible in the IGBT because it can switch on and off the main current with the MOS gate directly.
- In an on-state, its n-type channel formed by the MOS gate works as the cathodal n-region of the pin diode.
- The reduction of the on-voltage of the IGBT has been improved mainly by reducing the influence of the emitter-side p-region [23].

The IGBT's on-voltage had been so high that it took about 10 years from its commercialization in 1985 to surpass the power transistor (BJT). That was mainly because it did not deal with the third item in the list above and shortening the n^--region length was insufficient besides. For instance, the on-voltage of the IGBT was larger than that of the BJT until about the middle of the 1990s whereas the BJT of the 1200 V class consisted of a three-stage Darlington connection.[259] The reason why an appearance of the

---

[258] Refer to Fig. 2.113.
[259] It is indicated as the "1 kV transistor" in Fig. 1.4.

158 | *Basic Technologies of Major Power Devices*

IGBT was desired earnestly in spite of this was primarily that the driving power supply circuit became small.[260]

In the BJT, 1200 V was the practical limit to use, but various improvements were made in the IGBT and the product group up to 6.5 kV had been prepared. Although the IGBT chip becomes expensive because production equipment of the same level as that required for manufacturing a large-scale integrated circuit (LSI) is necessary, the IGBT was able to get competitive by reducing the chip size because it can handle about 1 order larger current than the BJT.

2.9.2 Basic Operation of the IGBT

As for the operating principle of the IGBT, there has been two kinds of interpretations from the beginning of development: the pnp transistor that is driven by the MOSFET and the series connection of the pin diode and the MOSFET. GE regarded the first IGBT trial product of 1982 as the series connection of the pin diode and the MOSFET and called it the insulated gate rectifier (IGR) [52]. However, the company soon began to illustrate that it was the composite type having a pnp transistor driven by a MOSFET. Toshiba, which manufactured the IGBT first, also explained the IGBT operation in this model.[261] Therefore, the model to drive the pnp transistor by the MOSFET had been more or less fixed as the theory of IGBT operation.[262]

However, this model was not able to give the guiding principle that was effective in improving its on-voltage, which was a great hindrance to putting the IGBT to practical use.[263] For this reason, the issue regarding whether the series connection of a pin diode

---

[260] After the middle of the 1980s when the inverter had begun to spread, there was severe competition among the general-purpose inverter makers to downsize and bring down the price. Then, downsizing of the base drive unit to supply about 1/50 of the main current was strongly demanded. The high-speed operating property particular to the IGBT was the secondary advantage. Such a situation can be seen because the inverter makers had not emphasized the replacement of the BJT by the IGBT at all.

[261] The company called it the bipolar mode MOSFET (BiFET).

[262] The inevitable reason to choose this model would be that an IGBT must be thought to work as a pnp structure device during the off operation.

[263] The latch-up, which made an IGBT uncontrollable, was a further problem.

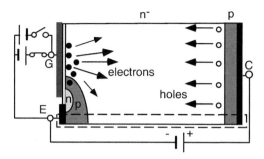

Figure 2.106 Schematic diagram of planar-IGBT operation.

and a MOSFET was appropriate to express the on-state of the IGBT appeared again in the beginning of 1990s[264] [11, 53].

In about 2010, it was stated that the IGBT operation did not satisfy the basic requirements of the BJT operation [35][265] and the device simulation result that reproduced the on-state operation of the IGBT, namely distributions of the charged carriers and the current densities, in the partial pin diode that changed a part of the p-region to the n-region [23]. It was then that the operating model of the IGBT, that is a series connection of the partial pin diode and the MOSFET, began to be finally accepted, after about 20 years from the trial manufacture of the IGBT [55].

In the on-state, both planar-IGBT and trench-IGBT are considered to be the parallel connection of the "MOSFET + i-region + p-region" and the pip structure between the collector and the emitter, as shown in Figs. 2.106 and 2.107. This "MOSFET + i-region + p-region" corresponds to the series connection of the MOSFET and the pin diode. It is because a channel of the MOSFET functions as the n-region of the pin diode. But this pin diode is partial at the point where the p-region also exists along with the n-region on the cathodal side.

---

[264] It can be considered that during the off-operation of the IGBT, the charged carriers, which were stored at the on-operation, are merely deducted from the i-region of the pip structure and it is not necessary to think about the pnp transistor operation.

[265] That is, either free electrons or holes consist of the collector current depending on npn-type or pnp-type, respectively, in the BJT, but both constitute the collector current in the IGBT.

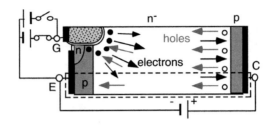

Figure 2.107 Schematic diagram of trench-IGBT operation.

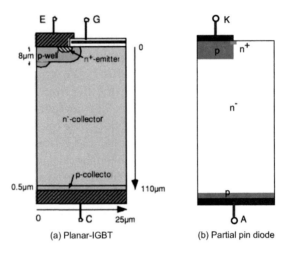

(a) Planar-IGBT   (b) Partial pin diode

Figure 2.108 Structures of the planar-IGBT and partial pin diode.

It can be confirmed that such simplification is effective from the comparison of the device simulations of the IGBT and the partial pin diode as shown in Fig. 2.108 [23].[266]

The left planar-IGBT is the NPT type having a thin p-region as the collector layer.[267] Device width W: 25 μm; n^--region: 110 μm long, 1×10^{14} cm^{-3}; collector p^--region: 0.5 μm thick, 10^{18} cm^{-3} (uniform); p-well: 16 μm wide; surface concentration of p-well:

---

[266] Those were the isothermal simulations neglecting the temperature rise by the operation at room temperature. A basic structure such as Fig. 2.108 is called a "cell."

[267] The structure of the emitter side is almost the second-generation structure of the early 1990s (refer to Fig. 2.97). To select this structure, the easiness of simulation was given priority over reproduction of the real structure.

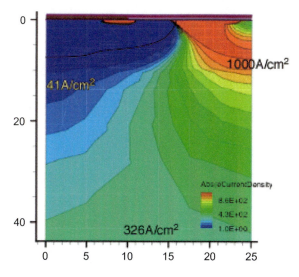

Figure 2.109 Free electron current distribution of the planar-IGBT. Reprinted from Ref. [57] with permission from the Institute of Electrical Engineers of Japan.

10^{19} cm^{-3} (Gaussian profile); $V_{th} \approx 4$ V; SRH-type lifetime (τ_e: 10 μs, τ_h: 3 μs); and gate oxide thickness: 50 nm.

As for the right structure of the partial pin diode, a fairly large p-region (10^{18} cm^{-3} uniform, 4 μm thick, and 8 μm wide) exists on the cathodal side along with the n$^+$-region (10^{19} cm^{-3}, 0.01 μm thick) that could be considered to play the role of the MOS channel. The projected length of the n$^+$-region $NLen$ from the p-region of the surface to the n$^-$-region is 2 μm. The device width W is 20 μm; the n$^-$-region is 100 μm, 1×10^{14} cm^{-3}; and the anode p-region is 0.5 μm and 10^{18} cm^{-3} (uniform). Recombinations of the charged carriers are neglected.

Figures 2.109 and 2.110 are distributions of the free electron current J_e and the hole current J_h in an area about 40 μm below the emitter surface in the planar-IGBT (at $V_{CE} = 5$ V, $V_{GE} = 10$ V) [57].

Figures 2.111 and 2.112 are distributions of the free electron current J_e and the hole current J_h in an area about 40 μm below the emitter surface in the partial pin diode (at $V_{AK} = 5$ V) [23]. Although these two figures are the result of the adjusted the partial pin diode

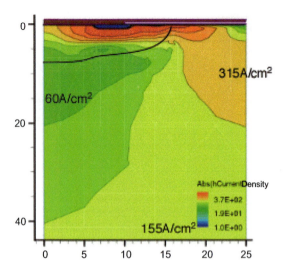

Figure 2.110 Hole current distribution of the planar-IGBT. Reprinted from Ref. [57] with permission from the Institute of Electrical Engineers of Japan.

structure[268] to reproduce the situation of Figs. 2.109 and 2.110, both structures result in similar current distributions.[269]

That is, the current distribution of the planar-IGBT can be almost reproduced in a partial pin diode whose cathodal structure has the n⁺-region projected a little, which can be considered as the MOS channel, from the surface of the p-region of the same electric potential. This supports the interpretation that the IGBT is the partial pin diode in which the p-region exists along with the n-region on the cathode side. Flows of free electrons and holes in the on-operation of such a partial pin diode are expressed as in Fig. 2.113. The free electrons flow out exclusively from the cathode n⁺-region. The holes that flow from the anode p-region are distributed in about the same density of these free electrons to keep an electric charge neutrality. The holes can also flow into both the n⁺-region of the

---

[268] It was premised that the thicknesses and impurities dose of the anode p-regions were common in both structures. Except for it, the most strongly influenced factor was the projected length $NLen$ of the surface n⁺-region from the p-region to the n⁻-region. There was little influence of the difference in the device width W, 25 μm and 20 μm, and the n⁻-region length, 110 μm and 100 μm.

[269] The figures of the IGBT and the diode classify current densities at equal intervals in logarithm indication. In addition, the values of J_e and J_h are also indicated near the corresponding points.

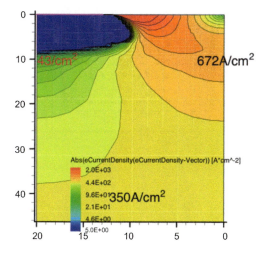

Figure 2.111 Free electron current distribution of the partial pin diode. Reprinted from Ref. [23] with permission from the Institute of Electrical Engineers of Japan.

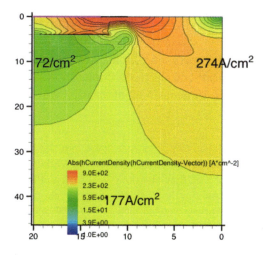

Figure 2.112 Hole current distribution of the partial pin diode. Reprinted from Ref. [23] with permission from the Institute of Electrical Engineers of Japan.

cathode side and the p-region. However, the hole current J_h flowing into the p-region becomes rather less because the free electron density n_e near the p-region of the cathode side is small.

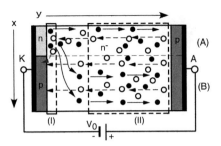

Figure 2.113 Carrier flows in the partial pin diode (●:free electron, ○:hole).

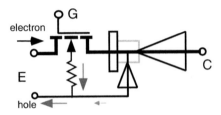

Figure 2.114 Equivalent circuit of the IGBT.

Considering these current flows and the connection between the n^+-region and the cathode electrode through a MOSFET channel in the real IGBT, the equivalent circuit of the IGBT would be expressed as in Fig. 2.114.

> In the partial diode, the holes that come into the n^+-region of the cathode side flow into the neighboring cathode electrode immediately, but they flow into the p-well close to the channel of the MOSFET portion in the IGBT. This hole current component is expressed by a current flowing through the resistance connected to the substrate of the MOSFET. In addition, the anode mark of diode "△" means the p-well of a vertical diode.

It would be apparent that an equivalent circuit of this IGBT is applicable to not only the planar-IGBT but also the trench-IGBT. Besides, the simulated IGBT mentioned above was the NPT-IGBT of Fig. 2.108a, which consists of the very shallow collector p-region without the n-buffer layer, namely the transparent collector. Even

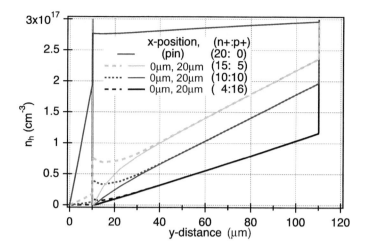

Figure 2.115 Dependency of hole densities on the n/p-area ratio (at $V_{AK} = 1$ V). The solid line and the broken line correspond to $x = 0$ μm and 20 μm, respectively.

for the collector structure of the PT-IGBT, a similar result must be obtained if adjusting the anode characteristic of the partial pin diode equally.

Figure 2.115 shows the distributions of the hole density n_h in the emitter-collector direction when the cathode n-region ratio in Fig. 2.113 is changed.[270] If there is even a small p-region in the emitter region, the hole density n_h of the n⁻-region falls toward the emitter side from the collector side linearly. The lowering degree from the hole density distribution of the pin diode is large so the ratio of the n-region decreases. Furthermore, charged carrier distributions are approximately uniform in the x direction, namely $n_h(0\ \mu m) \approx n_h(20\ \mu m)$, except the neighborhood of the emitter region.

This is mechanism to bring Fig. 2.102, which shows improvement of the charge carrier distribution by the IEGT or the CSTBT.[271]

---

[270] The free electron density n_e in the n⁻-region becomes the sum of this n_h and the impurities density N_{Dn^-} of the n⁻-region.

[271] It has been said that these new IGBT structures improve the charged carrier densities on the emitter side, but an effect to increase them, $n_h \approx n_e$, in the whole n⁻-region is much bigger.

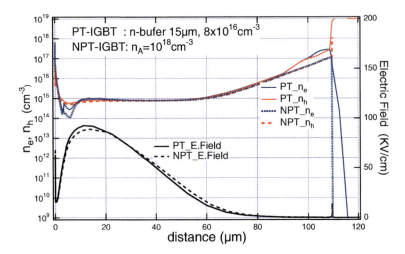

Figure 2.116 Distributions of EF, n_h, and n_e of a PT-IGBT and a NPT-IGBT in a short-circuit operation. ($V_{CE} = 380V$, $V_{GE} = 15V$, left: emitter, right: collector). Reprinted from Ref. [37] with permission from the Institute of Electrical Engineers of Japan.

2.9.3 High-Voltage Large-Current Operation of the IGBT

The typical on-operation of the IGBT in a high voltage is the short-circuit operation. Figure 2.116 is an example of simulated results of a PT-IGBT and a NPT-IGBT at $V_{CE} = 380$ V, $V_{GE} = 15$ V. Both IGBTs have the same emitter side structures and n$^-$-regions, and their collector side structures are adjusted so that their BV$_{CES}$ values are at the same level, ~600 V. Then, the internal distributions of both IGBTs become nearly the same [37]. The on-voltages of both IGBTs are also similar: NPT-IGBT = 2.0 V and PT-IGBT = 1.9 V at $J_C = 100$ A/cm^2.

> For the NPT-IGBT, the same structure as the previous subsection, namely Fig. 2.108a, was used. Although no lifetime control was done for the NPT-IGBT, its SRH-type lifetimes were selected as $\tau_e = 1$ μs and $\tau_h = 0.3$ μs as the middle values of both types of IGBTs to make the condition even with the PT-IGBT.
>
> For the PT-IGBT, the peak impurities density and the thickness of the n-buffer were 8×10^{16} cm^{-3} and 15 μm and those of the p$^+$-collector region were 1×10^{19} cm^{-3} and 10 μm. The SRH-type lifetimes were the same as those of the NPT-IGBT. $V_{th} \approx 4$ V.

The densities of the holes and the free electrons are approximately equal, $n_h \approx n_e$, and in the whole area of the n^--region, the EF is almost inversely proportional to the charge carrier densities n_h and n_e. From this, it is obvious that holes and free electrons move through the n^--region by the drift mechanism, $J_{drift} = qn\mu$ EF.

The IGBT belongs to the pin diode group the same as the GCT, but its distributions of the n_h, n_e, and EF are very different from the almost uniform ones of the GCT.[272] Figure 2.116 rather resembles the internal distribution of the short-circuit operation of the BJT, that is, Fig. 2.82. But the charged carrier densities in the high-electric-field region exhibit orders of magnitude difference. Although the densities of holes and free electrons in the BJT exceed only just the impurity density of the n^--region, about 10^{14} cm^{-3} order, in contrast, there are about 10^{15} cm^{-3} of those uniformly in the IGBT, as shown in Fig. 2.116.

In other words, the internal situation of the IGBT through which a large current holding a high voltage flows shows characteristic between the GCT and the BJT. That is, because the IGBT must be forced to operate as a partial pin diode because of the existence of the emitter-side p-region, namely the p-well, the GCT operates nearly as the pin diode.[273]

On the other hand, the IGBT in a turn-off process after its gate has closed can be expressed as in Fig. 2.117. An accumulation area of charged carriers is demolished by the extension of the electric field region W_E with the increase of the collector voltage V_{CE}. And the holes and the free electrons of the demolished portion consist of the emitter current J_{Eh} and the collector current J_{Ce}, respectively. Moreover, holes that are generated by impact ionizations are added to this emitter current J_E. Then the emitter current only consists of holes. In contrast, the collector current J_C comprises J_{Ce} caused by

---

[272] This is the situation of the pin diode where the n_h and n_e are smaller, as in Fig. 2.43.

[273] The GTO or the GCT operates as a partial pin diode too in comparison with a thyristor because it uses an anodal short structure to prevent the excessive accumulation of charged carriers in the n^--region during the on-operation. However, the emitter short structure of the IGBT has a much stronger effect because the interval between its p portions is very short.

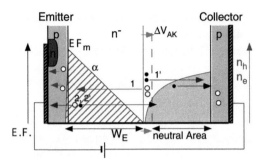

Figure 2.117 EF, n_h, and n_e of the IGBT during the turn-off process.

the free electrons from the demolished accumulation area and the hole current J_{Ch} from the collector p-region.[274]

Figure 2.117 resembles the turn-off process electric field and charge carrier distributions shown in Fig. 2.58, and only the shape of the n-region in the emitter surface is different. In the GCT, the n-region is formed widely on the emitter surface and an electrode is connected to its entire surface. On the other hand, the n-region of the IGBT has been formed as small as possible to prevent thyristor action, namely latch-up, which is only established as the path of the current.[275] In addition, the quantity of charged carriers remaining in the off-process in the IGBT is much smaller than that in the GCT. It is because the charged carrier densities in the on-state of the IGBT are several orders smaller than that of the GCT. Both Figs. 2.58 and 2.117 can be regarded to show the high-voltage operation of the pip structure devices where a large quantity of charged carriers are accumulated in their i-regions. And thyristor action begins immediately in Fig. 2.58 (GCT) when an impact ionization phenomenon happens in a high-electric-field area. In contrast, in Fig. 2.117 (IGBT), the operation of the pip structure continues unless a parasitic thyristor starts to work.

---

[274] When the hole that was demolished from the accumulation area moves to the emitter side, the electric potential of the neutral accumulation area outside the collector p-region decreases. Then the internal potential difference of the n^--p junction becomes small and free electrons and holes move to the opposite directions, stepping over this junction as in an on-operation.

[275] Figure 2.117 is conclusively different from Fig. 2.58 in that there is no flow of free electrons from the cathode n-region to the anode p-region.

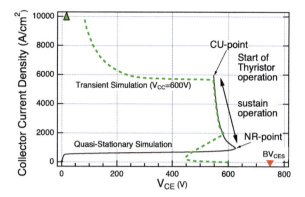

Figure 2.118 Basic loci of the IGBT's sustaining operation: ultimate simulation of the short-circuit operation ($t_{n^-} = 110$ μm).

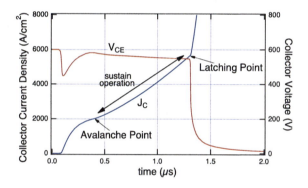

Figure 2.119 Example of transient simulation of the IGBT's short-circuit operation.

Figure 2.118 shows examples of device simulations on the short-circuit operation of the IGBT until the calculation limit. The solid line means a quasi-static simulation, and the broken line is a transient simulation.[276] At a large current density of more than 1 kA/cm², it became unrelated to the gate voltage V_{GE}. Although the quasi-static simulation terminated at the CU-point where $J_C \approx 6$ kA/cm², the transient simulation reached $V_{CE} \approx 13$ V, $J_C \approx 55$ kA/cm², after several microseconds. The thyristor action, latch-up, happens at the

---

[276] Figure 2.119 shows the J_C and V_{CE} at the time when V_{GE} was increased from 0 V to 10 V in 20 ns at $V_{CE} = 600$ V. Moreover, a huge built-in resistance was used in the quasi-static simulation.

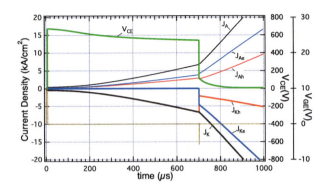

Figure 2.120 Example of the IGBT's destruction at the L-load turn-off.

CU-point. It was the IGBT structure of about 1993, however, and the latch-up current was improved phenomenally after that. That is to say, although the operating voltage of the IGBT is basically restricted by its own sustain voltage, its operating current would be substantially unlimited if its operating time was so short that a rising temperature could be ignored.

2.9.4 Safe Operating Area of the IGBT

The SOA of ideal IGBTs can be indicated with a sustain curve as shown in Fig. 2.118.[277] If the collector current J_C gets larger than the CU-point, where the thyristor action happens, it is obvious that it will lose the holding ability to withstand an applied voltage. If the collector voltage V_{CE} is larger than the value on the sustaining curve that corresponds to J_C, the J_C would continue to increase and lead to the thyristor action as shown in Fig. 2.120 when the sustain curve is negative resistive. On the other hand, when the V_{CE} is below that point, the operating point goes down the sustain curve with the decreasing electromotive force of the L-load and would gently finish the off-operation as shown in Fig. 2.121 if the temperature rise of the IGBT is not too large.

However, SOAs of real IGBTs are much smaller than expected from the simulation mentioned above. Its main reason is an

---

[277] It is basically the same as the SOA of a BJT, which is explained in Fig. 2.72.

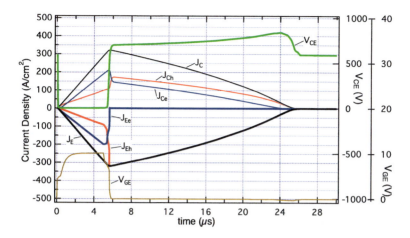

Figure 2.121 Example of the IGBT's L-load normal turn-off.

in-homogeneous operation in the chip [38]. Examples of actual IGBTs' SOAs and their simulation results are explained in the following.

2.9.4.1 Observation of IGBT destructions

Destructions during the short-circuit operation are classified into four kinds, as expressed in Fig. 2.122.

#A is the normal destruction waveform and is caused sooner or later by a temperature rise if the operating period becomes long. At that time, it is estimated that the p-n junction of the emitter side vanishes at a limit temperature of about 650 K [39].[278]

#C is also caused by a temperature rise, but an IGBT gets destroyed after a nearly normal turn-off as a measurement example of Fig. 2.124. During a short-circuit operation, the temperature of the IGBT chip rises adiabatically and the leakage current could continue its temperature in a positive feedback phenomenon even after the

---

[278]It was estimated as the temperature where the consumed energy until the destruction was extrapolated to a zero in the experiment where an ambient temperature was changed from 300 K to 450 K in Fig. 2.123. This extrapolation temperature, ≈650 K, was independent from the n^--length, lifetime control of the IGBT, and also the measuring voltage (V_{CE} = 600 V, 720 V). Furthermore, this basic mechanism was confirmed by a device simulation.

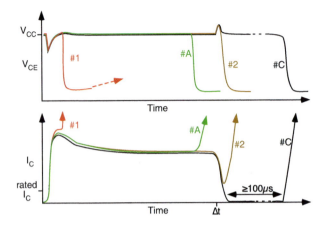

Figure 2.122 Short-circuit waveforms and destruction points.

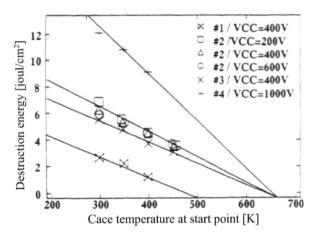

Figure 2.123 Destruction energy of #A type versus temperature. (#1 sample had an older emitter structure than others.) © [1996] IEEE. Reprinted with permission from Ref. [39].

short-circuit operation and might cause thermal runaway since its temperature does not decrease immediately[279] [40].

---

[279] Because the thermal resistance of the solder that die-bonds a chip is large, only a silicon chip would be heated up in the extremely short operation as the short-circuit operation. This tendency is common to all devices, but the temperature rise would become outstanding in FS-IGBTs because their chip is thin and their thermal capacity gets small.

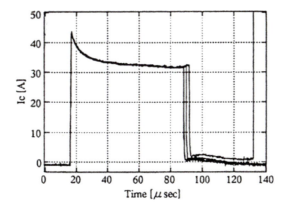

Figure 2.124 Destruction after turn-off of the IGBT: t_{ON} dependency. © [1994] IEEE. Reprinted with permission from Ref. [41].

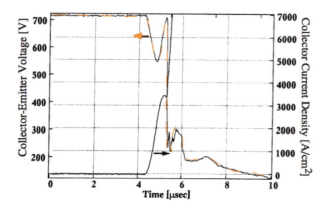

Figure 2.125 Example of #1-type destruction in an IGBT. (600 V IGBT, $V_{CE} = 720$ V, $V_{GE} = 50$ V, $R_G = 330\Omega$, at 300 K). © [1993] IEEE. Reprinted with permission from Ref. [42].

#1 is a typical destruction that occurs just after the beginning of the short-circuit operation, as shown in Fig. 2.125. It was the typical destruction mode in BJTs but extremely rare in IGBTs.[280]

#2 is the destruction at the off-timing of the short-circuit operation. At this time, it seems easy to be destroyed because of the

---

[280]It could be barely observed and that too in unreasonable conditions such as a power supply voltage of 600–720 V and $V_{GE} = 50$ V.

174 | *Basic Technologies of Major Power Devices*

V_{CE} surge, but it was almost not observed except in the early days of development.

On the other hand, the L-load off-operation can be divided into two kinds depending on the power supply voltage. One is unclamped inductive switching (UIS), where the power supply voltage is less than several dozens of volts and the load is a pure coil, namely it is typical to dive the ignition coil of the gasoline engine.[281] The IGBT was expected not to be destroyed instantly in the UIS test. It is because there is no worry that the transistor will be turned on in the IGBT unlike in the BJT, and the operating current of UIS, which drives a large L-load by the low power supply voltage, is not more than the rating current. However, it was reported in 2003 that an IGBT was destroyed right after the collector current of the rating level increased if its lifetime was much shortened, ≤ 0.2 μs [43].[282]

The other is the L-load off-operation of the high-power supply voltage represented by inverter equipment. Because an FWD is certainly connected in parallel to the L-load, the substantial L-load is a stray inductance L_S of the main line in a switching circuit. The L-value is smaller than the UIS operation, but the operating current can become huge since the power supply voltage is high.[283] The destruction occurred when the collector current I_C began to decrease just after the I_C arrived at the maximum point.

Table 2.4 shows the measured limit values of the IGBT's L-load operation. The limit power density did not reach 1 MW/cm², and it

---

[281] It is performed as a simple and easy evaluation examination of the destruction endurance capacity of the BJT for igniters and is called "avalanche proof level test." The process that the UIS operation leads to destroy BJTs begins when the generated holes by a sustain operation in the n⁻-region come into the base area and turn on the central part of the emitter that is the farthest from the base electrodes for a first. Then, if the V_{CE} value of an on-operation part was lower than the power supply voltage, the on-state continues and the temperature of that part becomes very high and leads to thermal runaway.

[282] Observed wave forms were like those of Fig. 2.129. When the lifetime of a high-speed IGBT was getting smaller, the destruction energy of UIS test suddenly changed from ≈ 1.5 J/cm² to ≤ 0.2 J/cm² at $V_{CE(sat)} = 2.7$ V, where $V_{CE} \approx 700$ V, $J_C \approx 200$ A/cm² (corresponding to the rating current), and $L \approx 75$ μH cm². That IGBT had a trench-gate and PT structure.

[283] Although both examinations use the same measurement circuit except the L-load, these cases should be distinguished by the voltage of the power supply, not by the L-value. Usually, the testing voltage might be more than half of the rating voltage for a high-voltage measurement and it might be about 100 V for the UIS test.

Table 2.4 Observed destruction values at L-load turn-off of IGBTs at 125°C

Year ref.	Dest. power (MW/cm^2)	J_C (A/cm^2)	V_{CE} (kV)	IGBT class
1998 [44]	0.72	200	3.6	4.5kV
	0.55	150	3.65	
2000 [45]	0.7	245	2.7	for DC
	0.1	30	4.2	2.8kV
2000 [46]	0.36	90	4.5	6.5kV
2000 [47]	0.44	95	4.6	4.5kV
	0.29	61	4.69	
2002 [48]	0.61	143	4.3	6.5kV

tended to become small rapidly when the power supply voltage V_{CC} was increased.[284]

2.9.4.2 Destruction mechanism of real IGBTs

It has been clearly evident that the IGBT has a much larger SOA than the BJT, but the actual ability is extremely small compared to the latent potential that was expected in the Section 2.9.3. To investigate its reason a device simulation had been tried to reproduce the observed waveforms using a circuit, as in Fig. 2.126, containing two IGBTs whose characteristics were slightly different. Those IGBT structures were based on the planar NPT-IGBT used in the simulation of Fig. 2.116.

Figure 2.127 shows the result of a transient simulation on the short-circuit operation in which a main IGBT-A and a small-area IGBT-B, whose projection portion in the p-well was deepened only by 0.5 μm more than the IGBT-A's, were connected in parallel as in Fig. 2.126. The ratio of both IGBTs' area was assigned to be 1:0, 7:1, 3:1, and 1:1 so that the total area became 2 cm^2.[285] The waveform

---

[284] It was said that IGBTs less than 1200 V had a "double square SOA," namely 400 A/cm^2 × 600 V > 0.24 MW/cm^2 for 600 V IGBTs, but there was no announcement of the concrete limit value of their high-voltage L-load off-operation.

[285] The simulation was done at room temperature in following conditions. The stray inductance: $L_S = 5$ nH, namely 10 nH cm^2, $V_{th} = 4$ V. The gate drive voltage was 10 V/0 V for on/off-state and both the rise and fall times were 20 ns. The gate resistances R_{GA} and R_{GB} were assigned to be inversely proportional to the area of each IGBT and their parallel sum was 5 Ω, namely 10 Ωcm^2. The SRH-type lifetimes: $\tau_e = 10$ μs and $\tau_h = 3$ μs.

176 | Basic Technologies of Major Power Devices

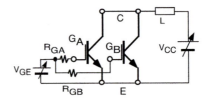

Figure 2.126 Simulated circuit of an IGBT's parallel operation.

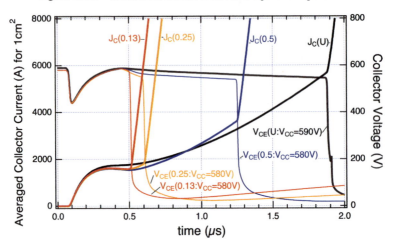

Figure 2.127 Area ratio of IGBT-B dependency of the whole J_C and V_{CE} (at $V_{CC} = 580$ V and 590 V). © [2002] IEEE. Reprinted with permission from Ref. [49].

of a uniform IGBT of area ratio 1:0, which is indicated as (U), is the same as in Fig. 2.119.[286] At the start of the sustained operation, current rises abruptly along with an increase of its flow area. Also, the observed waveform of this characteristic has been almost reproduced in the case of area ratio 7:1, as shown in Fig. 2.125.[287] At any area ratio, all currents finally flowed through IGBT-B and it was destroyed at the starting point of the thyristor operating in Fig. 2.118, namely ≈6 kA/cm² [49].

---

[286] The power supply voltage V_{CC} was 590 V only for this case, but for others V_{CC} was 580 V.

[287] It turned off normally after an on-period of 5 μs at $V_{CC} = 560$ V and it was destroyed at 570 V in an off-process and its waveform was similar to #2 in Fig. 2.122. Then it was destroyed just after turn-on at 580 V.

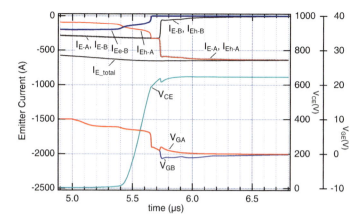

Figure 2.128 L-load off-operation of two identical IGBTs. L: 10 µH cm², 1 cm² × 2, V_{CC} = 600 V. Reprinted from Ref. [38] with permission from the Institute of Electrical Engineers of Japan.

Figure 2.128 is an example of the parallel off-operation of two identical IGBTs.[288] in the high-voltage L-load circuit. As a matter of course, both IGBTs showed the same operating waveforms at the beginning, but the whole current suddenly inclined to IGBT-A at 5.72 µs. Finally, all currents flowed through only one IGBT.[289] It suggests that all current would concentrate extremely easily in the portion in an IGBT chip exclusively where turning off was slightly difficult,[290] because the sustain operation is an intense positive feedback phenomenon for the current increase at the high operating voltage V_{CE}.[291]

---

[288] Those were equally set logically in a simulation procedure.

[289] Afterward, IGBT-A turned off successively at 42 µs after a sustain operation that was characterized by a gentle decrease of J_C and a slight increase of V_{CE}. By the way, the gate voltage difference ΔV_{GE} was not a cause of the current deflection because the gate voltages V_{GE} of both IGBTs were identical until 5.70 µs.

[290] Furthermore, a change of the gate voltage by the displacement current flowing in the gate resistance, which is caused by the change of the charge carrier densities below the gate electrode, accelerates the phenomenon. The delaying time of the off-operation got shorter by (i) the deeper depth of projection part of the p-well, (ii) the smaller value of V_{th}, and (iii) the longer lifetime. These influences seemed to be more dominant in the order of (i), (ii), and (iii).

[291] There was no particular difference between power supply voltages V_{CC} of 600 V and 100 V. However, it is certain that a higher V_{CC} can easily destroy an IGBT because a sustain waveform is not actually observed if the L-load or I_C is

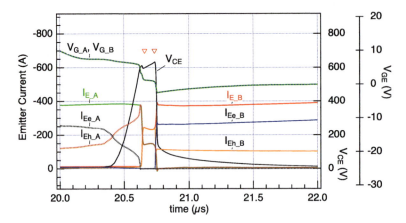

Figure 2.129 L-load operation of IGBT-A and IGBT-B with self-heating. A/B: 1/32, τ_A/τ_B: 1/1.33, $V_{CC} = 100$ V. Reprinted from Ref. [38] with permission from the Institute of Electrical Engineers of Japan.

In the L-load off-operation of parallel IGBTs, all sustaining current concentrated in one IGBT and lost voltage-holding ability only when the operating current density reached the starting point of the thyristor operation.[292]

However, the starting current at which the abnormal waveform happened got 25% smaller if containing the self-heating effect in the simulation[293] and the waveform got steeper as shown in Fig. 2.129 besides [38]. During 65 ns between the two ▽s in that figure, the temperature near the p-well increased by 86 K. According to it, the free electron current density J_e had begun to move to the outside from the projection part of the p-well, as shown in Fig. 2.130.

---

small. Moreover, the polarity of dI_C/dt at the time of the sustain operation for $V_{CC} = 600$ V is different than that for $V_{CC} = 100$ V, for instance, from the relation of $V_{CE} - V_{CC} = -L(dI/dt)$. Even so, it is because the internal conditions in the IGBT are almost the same since there is only a little difference between the I_Cs in those off-operations if using a large L-load.

[292] However, the simulated waveform at the destruction of the UIS operation in the power supply voltage $V_{CC} = 100$ V showed a gentler waveform than the observed one.

[293] The area ratio was A:B = 31:1. The L-load was 5 μH. IGBT-B's lifetime was 33% larger than IGBT-A's, which was $\tau_e = 10$ μs, $\tau_h = 3$ μs. Gate resistances were inversely proportional to the area of each IGBT and their parallel sum was 5 Ω. $V_{GE} = 10$ V/ 0 V for on/off-state.

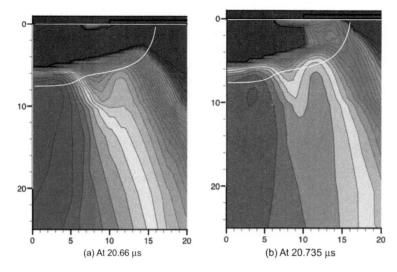

Figure 2.130 Distributions of the free electron current density at the L-load turn-off operation (1–9 kA/cm^2, in the log step). Reprinted from Ref. [38] with permission from the Institute of Electrical Engineers of Japan.

Just after that, a thyristor action happened and its V_{CE} decreased suddenly.

By the way, in a simulation at the second breakdown the self-heating effect has not been included sometimes, like the examples discussed in this book so far[294] because the phenomenon was so short that a rising temperature before the destruction would be negligible.[295] However, even if the conduction of heat needs time, the generation of heat does not have any time delay. If impact ionization happens locally, it raises the temperature at that point immediately.[296] Because an impact ionization action is depressed by

---

[294] It is also a good reason to take into account a lot of extra time if including self-heating.

[295] A semiconductor chip has much better specific thermal conductivity than a die-bonding solder. However, even if all the generated heat inside it was spent by the temperature rise of the whole chip, it is considered that the heating time would be extraordinarily slower than that of the second breakdown phenomenon.

[296] The heat is generated by the consumption of energy ($I \cdot V$). The impact ionization is a heat consumption phenomenon, rather.

a temperature rise, it would be necessary to consider the influence due to this self-heating sufficiently.

In addition, the contribution to thyristor action of the self-heating mentioned above is a phenomenon peculiar to the planar-IGBT and cannot apply to the trench-IGBT. It would seem to become important in trench-IGBTs that a displacement current through the gate oxide raises the gate voltage V_{GE}. Or the degradation of the gate oxide might occur by the current flowing through it.

Chapter 3

Applied Power Device Family: Power Modules and Intelligent Power Modules

Present-day equipment using power electronics has shown tremendous growth. This trend is expected to continue well into the far future as the demands for energy saving and alternate energy sources increase rapidly. To support the growth of power electronics, several evolutionary changes and breakthroughs have been achieved in the areas of power semiconductor device technologies. In the previous chapters physics related to various power semiconductors and their chip technologies have been described in detail. This chapter is devoted to providing details about applied power semiconductors' technologies, particularly those of silicon IGBT-based power modules and intelligent power modules (IPMs).

3.1 Review of the Power Module Concept and Evolution History

Figure 1.7, Chapter 1, explains the history of the whole power device family in a digest form. As explained before, the thyristor concept created the first wave and emerged as the solid-state

---

Power Devices for Efficient Energy Conversion
Gourab Majumdar and Ikunori Takata
Copyright © 2018 Pan Stanford Publishing Pte. Ltd.
ISBN 978-981-4774-18-5 (Hardcover), 978-1-351-26232-3 (eBook)
www.panstanford.com

182 | *Applied Power Device Family*

power switching solution for power conversion and triggered many possibilities for the growth of power electronics as a whole. As also highlighted, the second wave in the power device evolution history came along with the invention of the bipolar transistor in 1947. However, due to structural limitations of the basic transistor device proposed at that time, new innovations became essential to solving high-voltage/current-handling-capability-type switching components to simplify hurdles in designing high-power-rated conversion equipment and accelerate power electronics application growth.

Consequently, many research studies were carried out focusing on this need as power electronics applications became increasingly important to serve for saving energy, an awakening social consciousness that grew in the background of the global oil crisis in the early 1970s. The efforts in power semiconductor research under such forces eventually gave birth to a vertically structured bipolar junction transistor (BJT) that was suitable for voltage ratings above a few hundred volts. But the single-transistor npn-type vertical transistor element device that was initially created was not effective for high current rating. With further innovative approaches, a monolithically integrated npn Darlington transistor circuit, having two npn transistors vertically structured within the same silicon in a cascaded connection, brought in the real breakthrough and was used as the platform for designing high-power transistor chips and used in transistor power modules of voltage ratings ranging from 600 V to 1400 V and current ratings from 10 A to 400 A.

Figure 3.1 describes structural details of two typical high-power Darlington-type npn bipolar transistor chip architectures that were created for the bipolar transistor power modules of the 1980s. The circular chip architecture type shown on the left-hand side of the figure is a single-wafer-based design made for a very high current/voltage class. A unique vertical diffusion fabrication process technology was used to form a pair of monolithic integrated npn transistors in Darlington circuitry format (refer to the circuit topology given in the insert of Fig. 3.1). The bevel-type edge termination design together with the mesa chip concept is a unique solution used to achieve a high-breakdown-voltage characteristic. The transistor chip details shown on the right-hand side of Fig. 3.1

Review of the Power Module Concept and Evolution History | 183

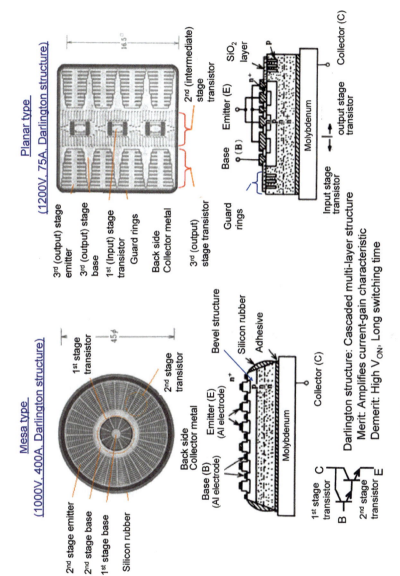

Figure 3.1 Structures of npn bipolar power transistors.

are another technological breakthrough achieved in the 1980s for the high-power-rating requirement. The chip architecture is based on a three-stage Darlington circuitry to achieve a very high current gain hFE (i.e., $h_{FE} = I_C/I_B \geq 100$, at rated $I_C = 75$ A current conduction) by a rectangular-type single-chip solution for a 1200 V class. The additional third stage over a typical two-stage Darlington circuitry indeed increased the collector-emitter saturation voltage drop, $V_{CE(sat)}$, of the device, giving higher on-state conduction losses, but that was an affordable choice for application designers. The rectangular chip is also shown to have guard rings at the surrounding edges, typically necessary for handling depletion layer build-up under a high reverse-biased voltage-blocking condition. Such unique high-power-rated bipolar transistor chip technologies made a breakthrough, giving birth to easy-to-use robust bipolar transistor power modules, which met application requirements adequately in the 1980s.

Although the bipolar transistor power modules (or bipolar junction transistor power modules or, simply, BJT power modules) spread into various inverter applications owing to an easy-to-use structural concept and high-power-handling capability, the application trend took different turns thereafter, arising from market and environmental needs.

Figure 3.2 shows the chronological trends of system needs and innovations in power device (module) seeds made to cope with such needs since the 1980s. Figure 3.3 summarizes evolutionary steps of major power module families. As revealed from the two figures, the metal-oxide-semiconductor (MOS)-gated fast switching type power transistors chip solution became an essential requirement for high-power inverter application from the late 1980s. The insulated gate bipolar transistor (IGBT) module thus became a key technology for power electronics applications. In addition to that, the intelligent power module (IPM) technology, which was first created and introduced by Mitsubishi Electric (Japan) in the late 1980s, became a superior alternative to any power switch solution available at that time due to its integrated intelligence and user-friendly features. The following sections will provide more details about these power modules' constructional and design aspects.

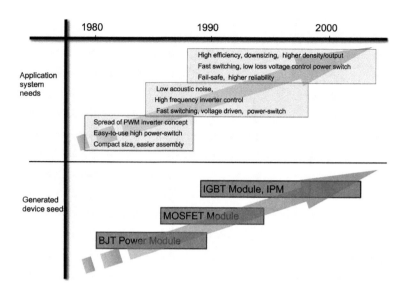

Figure 3.2 Flow of power module evolution coping with system needs.

3.2 Power Module Constructional Features and Design Aspects

The power module was first commercialized in the mid-1970s. Then, in the 1980s, the product family rapidly grew along with consensus building for energy saving. Generally speaking, a power module is formed by assembling several power chips and other passive/active components into a single housing or package. Its main use is power conversion and control. Before going into the details of power module features and functions, the following is a brief description about the power device family tree.

Figure 3.4 provides a broad classification of various power devices that have evolved so far and are being applied in different power electronics applications. The chart also gives an idea about rating ranges of standard devices of each category. Of these different types, the IGBT module and IPM families that are categorized under "Application-Specific Device" in the classification are dealt with in detail in the following sections of this chapter.

186 | *Applied Power Device Family*

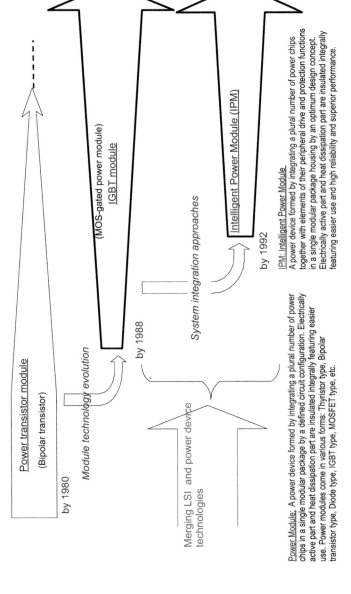

Figure 3.3 A summary of major power module evolutions.

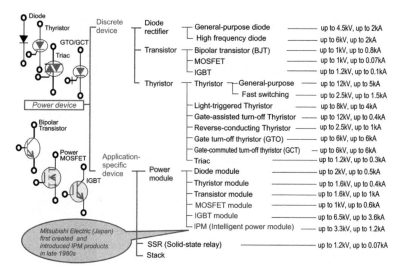

Figure 3.4 Power device classification.

Of the different types of power module categories shown in Fig. 3.4, the IGBT and IPM families are the most important current technologies. The typical power circuit configuration, or integrated functions, adopted in such power-transistor-integrated devices are shown in Fig. 3.5, taking the IGBT module family as an example.

As described in the figure, integrated circuit topologies are basically aimed at forming power circuit architectures in different inverter- and chopper-technology-based power conversion systems. Apart from the transistor/thyristor switch integrated modules like the old-generation bipolar transistor/thyristor types and the current-generation IGBT/metal-oxide-semiconductor field-effect transistor (MOSFET) types, the diode modules are designed mainly for rectification applications. They normally do not have any integrated switching element, but their internal circuit configurations are the same as those shown in Fig. 3.5. The IPM type, on the other hand, has more complicated integrated functionalities than those of the IGBT type or any other class, as briefly expressed in Fig. 3.4 and further detailed in Fig. 3.6. More details related to the IPM concept and features, evolution of different IPM families, and cutting-edge advances related to IPMs will be discussed in the later sections of this literature.

188 | Applied Power Device Family

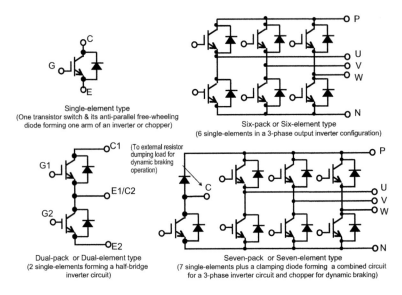

Figure 3.5 Internal circuit configurations of typical power modules (e.g., IGBT type).

IGBT Module: IGBT and diode power chips are integrated in multi-chip circuit formation in insulated module packages to meet some specific application purposes.
IPM: Integrates self-drive and self-protection functions in an IGBT module based on an optimized total design.

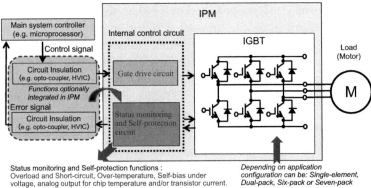

Figure 3.6 IGBT module and intelligent power module (IPM).

3.2.1 Basic Aspects of Power Module Construction and Design

On the basis of their structural features power module packages can be broadly categorized into two types: (i) case type and (ii) transfer-molded type. The first type uses an epoxy-resin-based insulated housing concept and is adopted for medium- to high-power-rated module families. The second one uses a specialized transfer molding encapsulation method and is adopted for designing low- to medium-power-range module families. The merits of and issues related to the two broadly classified classes are described in Fig. 3.7. The case-type structure, which is more abundantly available, can be further broken down into two subtypes, as shown in Fig. 3.8, to distinguish the integration of a control board for integrating intelligent functions as required for an IPM solution.

3.2.1.1 What are the characteristics required from a power module package?

As revealed by the earlier description, there are many structural elements besides power semiconductor chips in a power module that require careful concurrent engineering and expertise to ensure rugged casing design appropriate for mounting and connection requirements with adequate insulation and protection against environmental adversities. Besides exhibiting user-friendliness, a power module package is required to be good for forming specific electrical circuitry, with multiple devices having proper electrical insulation features and appropriate heat dissipation capability for the heat generated during the integral devices' operation, with adequate endurances against severe thermal and other environmental stresses. The casing of a module, irrespective of its category, provides reliable mechanical connectivity besides its user-friendly easy-adoption feature. For a typical case-type module, the enclosure is generally made of plastic (e.g., epoxy resin) material providing electrical insulation for electrically live components/parts inside. A casing having a silicon (Si) gel (silica) or epoxy resin filling performs as a basic protection against foreign particles and liquids. However, such case-type structures are generally not hermetically sealed. The

190 | Applied Power Device Family

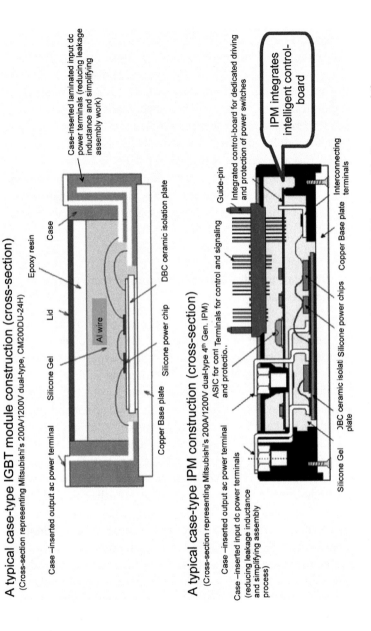

Figure 3.7 Comparison between case-type and transfer-molded (resin-molded)-type modules.

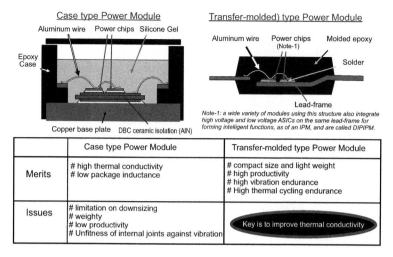

Figure 3.8 Comparison between two typical case-type devices: an IGBT module and an IPM.

packages are designed to have terminals achieving high-conductive electrical connections to the surrounding hardware.

Specifically, a power module package must satisfy the following requirements in addition to the standard properties of a semiconductor packaging concept, such as protection of integrated chips against mechanical stress and other environmental effects and simplicity of design.

- *Electrical isolation characteristic*: An appropriate electrical isolation feature is essential to cope with a high applied voltage.
- *Large current conduction capability*: Since a large current is conducted through internal electrical connections, conduction resistance and induction should be kept as small as possible to minimize heat generation and maximize the dynamic switching capability of the integrated power chips.
- *High conductivity of terminals*: Along with the large current conduction design aspects of the internal parts, the high-power-rated terminals have to be designed appropriately to provide a low resistance and reliable connections to their external hardware. On the other hand, terminals for control

circuit connections should be generally pin type, assuring a low contact resistance for low current conduction, irrespective of being soldering type or nonsoldering type.

- *Heat dissipation characteristics*: A power module generates heat caused by power losses in its ON-OFF operation cycle (i.e., conducting on-state and switching transient-state losses). Therefore, thermal design for dissipating the heat is important.
- *Endurance against thermal stress*: The temperature of a power chip inside a module can continuously swing rapidly from a low value to a high one due to heat generation during its operation cycles. Temperatures of other surrounding package parts can also swing repeatedly due to the similar thermal reason but at different temperatures and with different swinging ranges. Such disparities in temperature swings can cause alarming thermal stresses, and therefore, endurance against such stresses is an important aspect of module design.

3.2.1.2 What are the features and issues related to typical power module package designs?

In addition to the aforementioned description, the package structures of various power modules designed for various applications having different voltage and current ratings can be broadly classified into two categories: case type and transfer-molded type. First, the case-type structure will be taken up here, including its design features and constraints.

Figure 3.9 shows the cross section of a typical case-type IGBT power module having IGBT and diode power chips within its structure and mounted on an external heat sink with thermal grease applied in between two contact surfaces to reduce contact thermal resistance for the dissipated heat. Major design issues related to a module package design are also described in the figure. Figure 3.10 is a more detailed schematic of a high-power module housing giving additional details about the major integrated components in such a device. The power chips are soldered onto the isolation plates. For case-type power modules the classification of typical isolation plates is shown in Table 3.1. The first one in the table is a classic type used formerly for housing bipolar transistor modules. Due to its

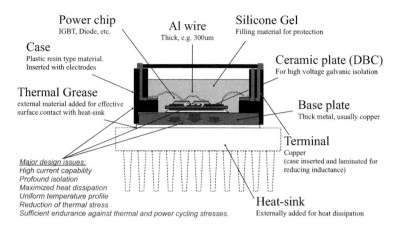

Figure 3.9 A typical case-type module package and related design issues.

Table 3.1 Isolation base plate structures for a case-type power module

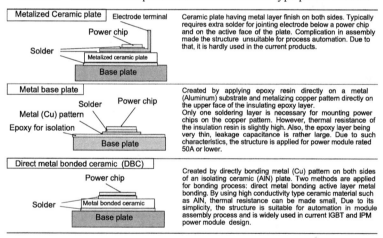

lack of an automation feature and a higher assembly cost, this type is basically not used in today's devices. However, this structure is regarded as a pioneer in the evolution of power modules. The second structure is a very cost-effective solution by virtue of its low material cost and simplicity in assembling. However, it is good only for the lower range of the current rating class due to thermal resistance and leakage constraints as described in the table. The third type is most

194 | *Applied Power Device Family*

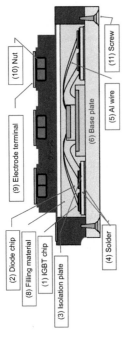

	Parts name	Main material	Main function	Electrical Characteristic	Life	Cost	Safety & environment conformity
(1)	IGBT chip	Si	Power switching and current conduction	○	○	○	
(2)	Diode chip	Si	Reverse conduction involving power switching	○	○	○	
(3)	Isolation plate	Ceramic/Cu	Current conduction, Heat conduction, mechanical jointing and electrical isolation	○	○	○	○
(4)	Solder	Pb/Sn→Sn/Ag/Cu	Current conduction, Heat conduction and mechanical jointing	○	○		○
(5)	Aluminum wire	Al	Current conduction	○	○		
(6)	Base plate	Cu	Body, Heat dissipation		○	○	
(7)	Case	Hard resin	Body, terminal encapsulation and electrical isolation			○	○
(8)	Filling material	Silicon, Epoxy	Electrical isolation				
(9)	Electrode terminal	Cu	Current conduction	○			
(10)	Nut	Cu (hexavalent chromium→Trivalent chromium)	Mechanical jointing	○			○
(11)	Screw						

Figure 3.10 A typical high-power IGBT module's internal structure and main integrated parts.

widely used for medium- to high-current-rated (75 A $\geq$) IGBT power modules and IPMs.

A ceramic isolation plate is formed by sandwiching a ceramic layer (typically <1 mm in thickness and made of aluminum nitride [AlN] or Al_2O_3 material) by copper layers. The chip and specially designed ceramic isolation plate, which is also known as the direct bonded copper (DBC) plate, is soldered onto a base plate (typically <5 mm and made of copper) and encapsulated by a specifically designed case structure typically made of a hard epoxy material and filling material such as silicon gel, also explained before. In a typical DBC structure, a copper layer is bonded directly to the ceramic sandwiched layer. The upper copper layer is the current-conducting layer patterned according to circuit requirements. This layer also acts to spread a portion of the generated heat inside the module. The lower layer is a nonconducting plain acting basically to spread heat and enhancing thermal-cycling endurance of the structure. Bonding the copper layers directly to the ceramic layer reduces the coefficient of linear thermal expansion (CTE) of copper, allowing a better CTE match between silicon chips and copper. Otherwise, the large thermal expansion of copper would cause crack formation and/or degradation of the solder layer because of the high-thermal-tension-related fatigue during temperature cycling. A wide copper layer can also cause ruptures on the ceramic because of a high tension generated by differences in CTE values. Table 3.2 shows

Table 3.2 Physical properties and features of various ceramic isolation materials

Isolation ceramic	Feature	Thermal conductivity (°C/W)	Resistive Strength (Mpa)	Insulation/semi-conductivity (Ω·cm)	Cost-effectiveness
Alumina Al_2O_3	The most widely used ceramic material. Good mechanical strength, electrical insulation, high-frequency loss, thermal conductivity, and resistance to heat, wear and corrosion.	20	350	$>10^{14}$	high
High-grade Alumina Al_2O_3		33	400	$>10^{14}$	medium
Aluminum nitride AlN	With its excellent electrical insulation and thermal conductivity properties, AlN is ideal for applications requiring heat dissipation. In addition, it has a coefficient of thermal expansion (CTE) near that of silicon, and excellent plasma resistance.	170	350	$>10^{14}$	low
Silicon nitride Si_3N_4	Silicon nitride is optimal for applications requiring toughness at high temperatures, superior thermal shock resistance, light weight and corrosion resistance.	80	600	$>10_{14}$	low

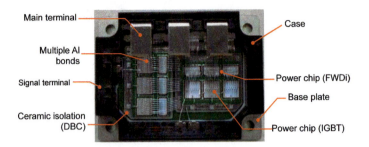

Figure 3.11 A picture showing the internal structure of a current IGBT module.

some key physical properties and features of different ceramic isolation materials that are used in the DBC structure explained earlier.

The copper material on the chip-mounted side of the DBC is designed has a specific pattern to make the electrical connection according to the circuit requirements. Ultrasonic wire bonding is used to make the connection between chip-to-copper-pattern and copper-pattern-to-terminal electrodes inserted into the case sidewalls.

Figure 3.11 is a photograph showing the internal structure of a typical IGBT module currently in use. This module is a half-bridge configuration and has secured the rated current capacity by paralleling three large-current-type IGBT chips and three antiparallel freewheeling diode (FWD) chips, also of a high current type. The structure is very close to the basic form shown in Figs. 3.5 and 3.6 and addresses most of the key design issues explained earlier. Using such modularized structural concept, it has been possible to rate power modules for high current conduction.

In Fig. 3.8 the considerable isolation (insulation) distances in a typical power module package structural design are shown numbered from (i) to (v). For providing an adequate isolation distance, it is necessary to consider both the creepage distance and the space (air) separation distance. For instance, each numbered distance in Fig. 3.8 will need to be designed following guidelines given in applicable International Electrotechnical Commission (IEC) and/or other standards.

External distances:

(i) Space (air) separation distance
(ii) Creepage distance

Internal distances:

(i) Distance between electrodes
(ii) Distance between two nearby electrodes at the chip-mounting surface
(iii) Distance between the active side pattern on the ceramic isolation plate and the metallic-based plate below the structure

It should be noted that the space/distance required for distances (iv) and (v) are much shorter compared to those for distances (i) to (iii) because the former group relates to internal distances where the whole space is filled with silicon gel and/or epoxy resin as the filling material and are inherently protected.

Figure 3.10 highlights typical design guidelines followed for achieving the desired isolation voltage and current conduction capabilities for building a standard DBC ceramic-based case-type power module.

3.2.2 Fundamentals of Power Module Structural Reliability and Life Endurance

Having described some of the key design issues, we will now examine structural reliability and life endurance aspects of a typical power module. Every power module has a definite life period [50–52]. In a typical power module, the integrated silicon power chips exhibit much longer individual lives than those of their structural parts. Several well-known lifetime models for predicting a module's life endurance have been presented so far. In general, two distinguished approaches can be outlined: analytical and physical. In any approach on such an analysis, a thorough investigation of the thermomechanical behavior of each part in the module assembly and coupling interactions among them are important. The following will briefly explain the two generalized life prediction models.

- *Analytical life endurance model*: This model describes the effect of temperature cycling parameters, for example, temperature

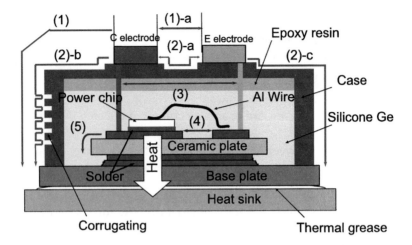

Figure 3.12 Module package design constraints for electrical isolation.

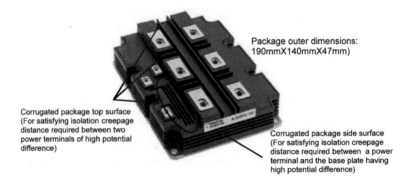

Figure 3.13 A 6.5 kV rated HVIGBT module package showing corrugated surface design features, needed for 10.2 kV isolation voltage specification.

variation/amplitude, duration of each temperature condition, frequency of the temperature cycle, and average/mean and highest/lowest temperature values, on the number of cycles to failure N_f. Among the several models given in the literature, a widely used one is the Coffin–Manson model, given below.

$$N_f = A(\Delta T_j)^{-n} \exp^{E_a/kT_m} \qquad (3.1)$$

Here, ΔT_j is the change in the junction temperature of a chip inside the module per cycle and T_m is the average temperature of the chip junction. According to this model,

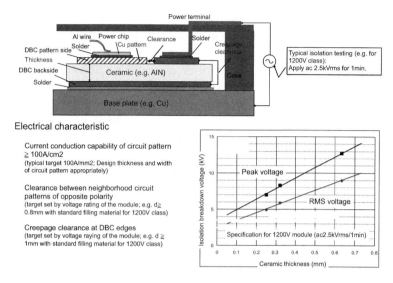

Figure 3.14 Power module constructional design aspects related to internal isolation structure and electrical characteristics.

the number of cycles to failure depends on the temperature variation ΔT_j and the average temperature of the power chip junction inside the module under consideration. This is a rather simple assumption in predicting the life of a complicated structure like a power module as it was found that other parameters, such as frequency of stress change (temperature cycles) and duration of heating and cooling times, also influence lifetime significantly. Norris and Landzberg considered these parameters and proposed a new analytical model for the number of cycles to failure, modifying the expression in Eq. 3.1, is given below:

$$N_f' = f^{-n2} N_f \quad (3.2)$$

Other multi-parameter-based models have also been proposed, taking new parameters into account, such as maximum junction temperature T_j, value of direct current (DC), I_{DC}, the module is expected to conduct, and other sets of curve-fitting constants extracted from long-term reliability evaluation experiments based on manufacturers' experiences.

- *Physical life endurance model*: This is based on the understanding of the stress/strain deformations within modules obtained either by experiments or by simulations. For the purpose, high-resolution measuring methods, such as infrared microscopy, and/or computation techniques, such as finite element analysis (FEA), are employed. The physical models for estimating thermal-fatigue-related lifetime of wire-bonding materials and the strength of chip-to-wire bonded contacts are based on fracture mechanism, and simulations by FEA are widely used for estimating the endurance limits. Physical models for predicting endurance limits of solder layers/joints in base structures of power modules can be based on four methods: energy based, damage based, stress based, and strain based. Out of these, the energy-based method is often employed because of convenience of usage and aptness of predicting actual conditions the module class being investigated would experience.

Many researchers have worked on the energy-based approach for predicting life endurance of solder layers in modules. M. Ciappa et al. proposed models for high-power devices in 2003. In this approach, it is assumed that the end of life of a device is based on the total deformation energy accumulated during the operation of a device and a device fails when the total deformation energy reaches a critical value, ΔW_{tot}. This model takes only time-dependent creep into account, neglecting elastic and plastic deformations. Many other researchers also worked on physical life endurance models, taking these deformations of solder layers caused by fatigue into account. One such new physical model for life endurance estimation of power modules was published in IPEC 2010. On the basis of the information earlier and referring to Fig. 3.15, it can be understood that the life period is mainly limited by the thermal fatigue of the jointing regions caused by any one or a combination of the following two main items:

- Wire junction fatigue from "power cycling"
- Solder junction fatigue from "thermal cycling"

The terms "power cycling" and "thermal cycling," expressing the stresses that a power module is subjected to in a typical application, are related to its operating conditions as explained in Fig. 3.16. The

Power Module Constructional Features and Design Aspects | 201

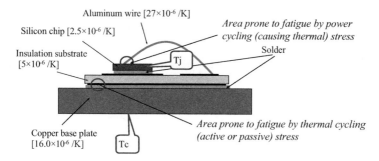

Figure 3.15 Life period consideration for a power module. (The values in brackets are coefficients of linear thermal expansion.)

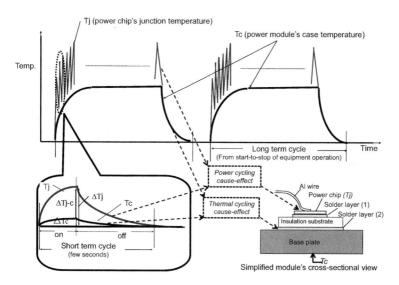

Figure 3.16 Typical cyclic variation of power module temperatures in an application making cause-effect stress to its structure.

cyclic variation of temperatures at a heat-dissipating power chip's junction with respect to surrounding conditions leads to specific stresses as defined by the operation and cause-effect model shown in Fig. 3.16. The stresses coupled with the CTE of different structural materials lead to thermal-fatigue-related damages that define the endurance limit of the module structure and, hence, its life.

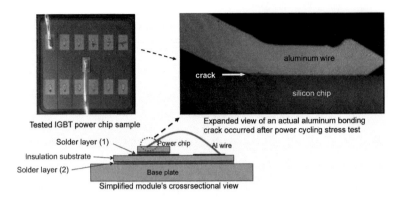

Figure 3.17 Typical example of a power cycling endurance test failure.

The following analysis provides details explaining the typical failure modes of case-type power modules caused by the two broadly classified thermal fatigue root causes discussed earlier. Figure 3.17 shows a typical failure pattern after a power cycling stress test performed using a standard case-type module. The power cycling stress applied in such a test is basically a rapid on-off current conduction through the power chips of the module under test (DUT), swinging the chip's junction temperature T_j by a predetermined and control value (e.g., $\Delta T_j = 50°C, 80°C$, etc.). As revealed, failure occurs at a wire-bonding contact on the associated chip surface. The crack in the bonding surface progresses by stress arising due to different rates of linear expansion (CTE differences) between the aluminum wire and the silicon chip and finally leads to an aluminum-bonding wire-peeling failure mode.

Different from the result of power cycling failure, a typical thermal cycling failure occurs by crack formation in the solder layer below the insulation substrate due to longer-type periodic temperature swinging stress, as described earlier and illustrated in Fig. 3.16. Figure 3.18 shows a typical failure pattern after a thermal cycling test performed on a typical case-type power module.

The stress strain in the solder layer between the insulation substrate and the copper base plate occurs from the difference of coefficients of linear expansion. If this stress is repeated, a crack starts forming in the solder layer (2) visible in Fig. 3.18. And if this

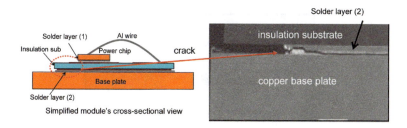

Figure 3.18 Typical example of thermal cycling endurance test failure.

crack extends to reach the part of the solder under the power chip by additional thermal fatigue stress, thermal resistance of the whole structure increases and the power module is very likely to get forced to a thermal runaway condition. However, if under such a state, the associated power chips under stressed condition experience a high junction temperature change or swinging (e.g., ΔT_j increases), forcing the assembly to cross its power cycling capability, as discussed earlier, the failure mode can finally transform to a wire peeling mode similar to the one shown in Fig. 3.17.

On the basis of the earlier discussion, it can be summarized that a typical case-type power module's operational life is limited by two key fatigue curves relating variation in temperature with the number of operation cycles—wire fatigue curve and solder fatigue curve—as shown in Fig. 3.17. The design considerations that are

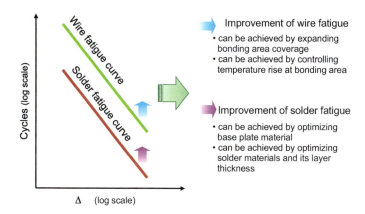

Figure 3.19 Fatigue curves.

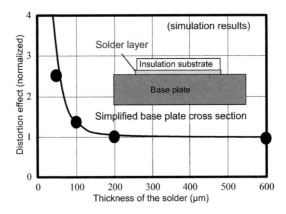

Figure 3.20 Solder layer thickness and distortion in the solder layer.

important for enhancing such module's fatigue endurance capability, or operation life, are also mentioned in Fig. 3.15. As mentioned in the figure, the solder layer in a typical base plate assembly structure plays a key role in improving the thermal cycle operation life of a typical power module.

The solder material to be chosen for the structure should basically be Restriction of Hazardous Substances (RoHS)-compatible (lead-free material) to satisfy environmental requirements and should also have the following characteristics:

- Good self-firmness and a low sliding tendency
- Good thermal conductivity
- Proper wettability to reduce void generation
- A CTE properly matched with that of the base plate as well as the insulation plate

As mentioned, the thickness of this solder layer is also important as its distortion effect limits the structure's endurance. Figure 3.20 shows a simulation result in this respect. In other words, a sufficiently thick and uniform solder layer should be constructed across the whole area, covering the insulation material and the base plate overlap in the structure.

The earlier description provides details regarding key constructional design aspects of a power module and factors determining its package structure reliability. The application designers on the other hand should consider many aspects and conditions under which the chosen module device would have to operate while meeting proper reliability and quality targets. Some questions that a designer must search for solutions at the beginning of his or her application design in this respect are given later.

- Application environment for a power module:
  - Does any environmental condition (ambient temperature range and its changing rate, humidity range and its changing rate, altitude, etc.) challenge the module's capability?
  - What is the intended thermal management—water cooling, a cooling temperature, or something else?
  - What is the estimated average and maximum junction temperature—high or low?
  - What is the estimated average and maximum junction temperature—high or low?
  - What is the worst-case estimation for mechanical vibration?

- Life expectancy for the power module:
  - How long should be the nonoperational life (storage, stand-by, nonoperational mode, etc.)?

Table 3.3 Typical test items and conditions applied in evaluating reliability of standard power modules for industrial applications

Category	Item	Test conditions	Related Std.
Environmental tests	Thermal shock	0°C 5 min +100°C 5 min 30 cycles	IEC 68-214
	Temperature cycling	-40°C 60 min +125°C 60 min 100 cycles	IEC 68-2-14
	Vibration	f=10 500Hz/15 min. 98m/s² (10G), 2 hours, X,Y,Z	IEC 68-2-6
	Tightening strength	Rating force	–
Endurance tests	High temperature storage	Ta=125°C 1000 hours	IEC 68-2-2
	Low temperature storage	Ta=-40°C 1000 hours	IEC 68-2-1
	Temperature humidity	Ta=60°C 90%RH 1000 hours	IEC68-2-3
	High temperature bias	V_{CE}= Rating voltage x0.85V Ta=125°C 1000 hours	–
	High temperature gate bias	V_{GE}= 20V V_{CE}= 0V Ta=125°C 1000 hours	–
	Intermittent current flow life	Tc=50°C 5000 cycles	–
Limit finding tests	Power cycle	Tj	–
	Thermal cycle	Tc	–

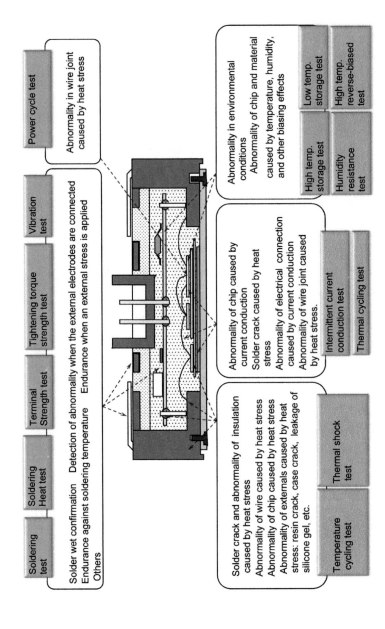

Figure 3.21 Summarizing reliability issues and examination methods related to power module construction.

- Is the needed operational life by passive temperature cycling (temperature swing at the chip junction/module case versus the number of cycles) well within the module's reliability characteristics?
- Is the operational life by active power/temperature cycling (active temperature swing at the chip junction/module case versus the number of cycles) well within the module's reliability characteristic?

3.3 State-of-the-Art Key Power Module Components

In this section, we will describe features of some of the latest power modules being widely applied today. First of all, we will look at the general trend followed currently in power module development. Figure 3.22 shows a perspective of the future trend that appears to be extending along the evolution path created in the 1980s. In other words, one major branch of evolution is toward higher integration, implying system-like integration, sustaining growth on the footsteps

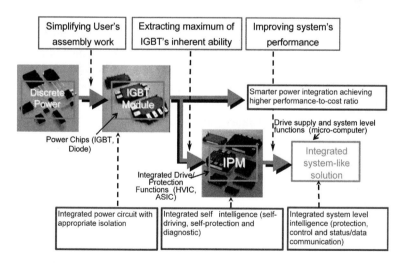

Figure 3.22 Major trends in power module technology.

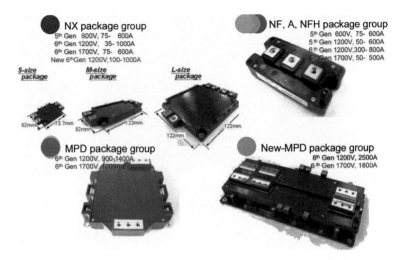

Figure 3.23 State-of-the-art IGBT module family (from Mitsubishi Electric's product portfolio, 2015).

opened by the IPM concept in the late 1980s. The other branch of sustaining progress is expected to follow the trend primarily toward smarter integration of power chips only and not much of peripheral intelligent functionality except for attaching mechanical features such as a heat sink.

Figures 3.23 and 3.24 describe some of the latest power modules from Mitsubishi Electric's current (2015) power semiconductor product portfolio offered for various power conversion applications. In this section, some of the state-of-the-art key power module component families, such as dual-in-line intelligent power module (DIPIPM) and up-to-date latest IGBT modules, are introduced.

3.3.1 Dual-in-Line Intelligent Power Module

In this section a detailed description of the state-of-the-art DIPIPM family, which is based on transfer-molding package construction technology, has been made explaining their features such as excellent productivity and reliability.

The DIPIPM product group—which became indispensable particularly to the inverter technology for home appliances, such

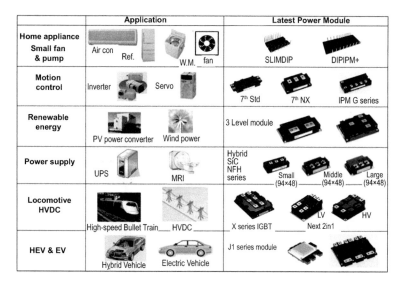

Figure 3.24 Latest module for each application (from Mitsubishi Electric's product portfolio, 2015).

as air conditioners, refrigerators, and washing machines, whose consumers have become increasingly conscious about energy saving and low-carbon environment issues—has contributed extensively to support such needs through its unique concept of the so-called all-silicon solution as shown in Fig. 3.25.

Figure 3.25 describes the original version of this breakthrough technology. The concept used in this technology was to have a smartly featured three-phase power module based on the latest power chip technologies (IGBT and diode) for inverter power conversion function and a dedicated drive control and protection interfacing based on newly developed high-voltage integrated circuit (HVIC) and application-specific integrated circuit (ASIC) technologies and all these functionalities integrated using a high-volume, productivity-oriented, transfer-molded compact packaging structure, as mentioned earlier. The all-silicon-solution concept-based product series soon became a de facto standard for low-power-end motor control applications, starting with the inverter-driven home appliances mentioned earlier. Keeping the essential features of the original concept, many technological innovations

210 | *Applied Power Device Family*

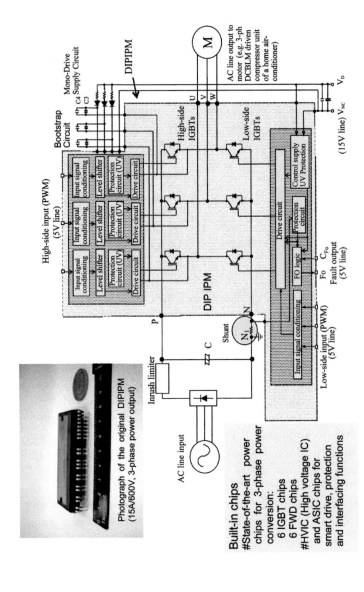

Figure 3.25 The original DIPIPM and its application circuit block diagram. Reprinted from Ref. [83] with permission from the Institute of Electrical Engineers of Japan.

State-of-the-Art Key Power Module Components | 211

Features	Circuit

• Integrate suitable power chips for application
  Low loss CSTBT, MOSFET..
• Low thermal resistance
  by Insulation resin sheet structure
• High current rating relative to package size
• Various application
  Photovoltaic generation
  PFC application (Full SW circuit + Rectifier)
	• DIPIPM standard topology series: 3 Inverter
	• *DIPIPM+ series*: Converter + Inverter + Break
	• DIPPFC series: Full SW circuit (PFC) + Rectifier

Functions
- Short circuit protection
- Control supply under voltage protection
- Over temperature (OT) protection
- Fault signal output in case of a failure
- Analog output of LVIC temperature (VOT)

Line-up of commercialized components

SLIMDIP	Super Mini DIPIPM	Mini DIPIPM / DIPPFC	Large DIPIPM	DIPIPM+
			50,75A/600V	50A/600V
+SLIMDIP-L/-S, 600V	5~35A / 600V	5~50A/600V	5~50A /1200V	5~35A /1200V
19×32×3.6[mm]	38×24×3.5[mm]	52.5×31×5.6[mm]	79×31×8[mm]	85×34×5.7[mm]

Figure 3.26 A summary of the latest DIPIPM products, with their features and functions, offered by Mitsubishi Electric (2015).

expanded the family to an extensive line-up covering a wide power-handling range (device ratings: 600 to 1200 V, 5 to 75 A) and enhanced functional features. Figure 3.26 shows a summary of the latest devices of this family offered by Mitsubishi Electric. A further description of this device group will be given in the following section.

3.3.2 Intelligent Power Module

As the trend in almost all power electronics applications continued to be toward higher efficiency, more ruggedness against abnormalities, steadier performance, higher noise immunity, higher operation frequency, downsizing, etc., focusing on inverter topology as the key technology, IGBT modules needed new dimensions toward higher functionality and better performance. To overcome the IGBT's inherent parasitic limitations and maximize its performance, fulfilling the ever-increasing application systems' needs, a concept of integrating dedicated drive and protection circuitry with the power element in an appropriate packaging was introduced by Mitsubishi Electric in the late 1980s based on a new invention. The resulting

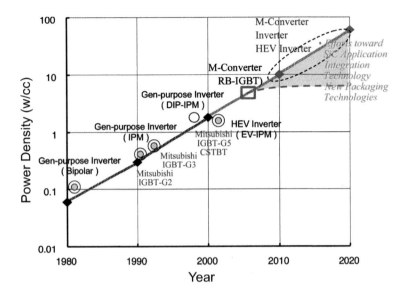

Figure 3.27 Projected growth of power density in power electronics system designs. Reprinted from Ref. [83] with permission from the Institute of Electrical Engineers of Japan.

module was named "intelligent power module," or IPM, because of its self-driving and self-protection capabilities.

As shown in Fig. 3.27, the power density factor related to power electronics equipment designs has improved remarkably in the past two decades. The main contribution in this growth came from using new power device components based on IGBT chip technologies, including different generations of IPMs. In the beginning, IPMs were applied only in low-volume and specialized categories of industrial drive applications.

In a few years after its debut, followed by the introduction of the revolutionary DIPIPM series, application of this unique product family expanded rapidly, starting with the high-volume zone of inverter-controlled home appliances. This primarily included air conditioners, refrigerators, and washing machines. The trend to apply an IPM instead of an IGBT was adopted by designers of other systems and, from their needs, various new versions of IPMs were developed and applied in different power electronics equipment,

State-of-the-Art Key Power Module Components | 213

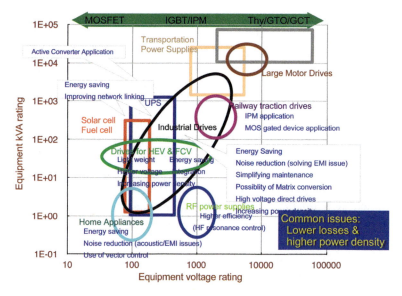

Figure 3.28 Application map of different power electronics equipment and their major technological needs. Reprinted from Ref. [83] with permission from the Institute of Electrical Engineers of Japan.

covering industrial motor controls, traction drive and auxiliaries installed in hybrid vehicles and railway systems, windmill power generation systems, solar power generation systems, etc. Through years of field experiences, the IPM has successfully shown its ability to perform adequately in a very wide range of power conversion applications (Fig. 3.28).

3.3.2.1 A review of the IPM's fundamental concept

The IGBT was first proposed in 1982, promising performance as an ideal power semiconductor switch by combining the best features of bipolar transistors and power MOSFETs [55, 80]. Recently, the IGBT became the main power device in the field of power electronics. Compared to a bipolar transistor and a MOSFET, an IGBT, through its two decades of refinement, exhibits a superior performance even at high voltage and high current and requires very low driving power.

However, every IGBT structure contains a parasitic thyristor element that further breaks down into two subelements, namely a parasitic npn transistor and a parasitic pnp transistor. These elements act adversely depending on the current density, the rate of change of current and voltage, and the temperature of the device. For instance, when a large current under a short-circuit condition flows through the device, the high current density condition can trigger the parasitic thyristor to turn on depending on the imposed stress level or duration of the same and the device temperature. When the parasitic thyristor latches on, the main IGBT's gate controllability is lost. Under such an uncontrollable state, system protection becomes a very difficult design task as within a few microseconds the device can get destroyed. This category of IGBT destructions is typically known as parasitic latch-up destruction. The parasitic thyristor's latch-up can also happen dynamically when the rate of voltage rise across an IGBT goes very high during its turn-off action or even at its static off-state. In such cases too, the parasitic thyristor's latching-on action becomes the root cause of the device's operation failure. In addition, the device temperature affects all such sustainable stress levels. The IGBT's capabilities to sustain such stresses without failure determines its safe operating area (SOA), which is an extremely important characterization needed for proper application designs. Enhancing IGBT's short-circuit-withstanding ability, and, hence, enlarging its SOA, has been the most difficult task all through for device engineers because the SOA capability bears a strong trade-off relation with the forward-voltage drop, $V_{CE(sat)}$, characteristic of the device. On the other hand, $V_{CE(sat)}$ characteristic bears a second trade-off relation with the turn-off speed and, consequently, turn-off switching energy of any IGBT structure. Thus a triangular trade-off relationship existing among three major characteristics complicates the IGBT's design process. The IGBT's design optimization in this respect has been the toughest challenge in the past three decades of its evolution. In the case of a normal IGBT module, the power elements inside are designed to sustain by themselves a certain amount of short-circuit stress, guaranteeing enough time duration and voltage-current safe operation area for external protection schemes to be designed appropriately in an actual application. The device is thus made rugged enough at the cost of the aforementioned forward-voltage drop performance and, hence, by worsening the power loss characteristic. Motivation to solve the inherent parasitic

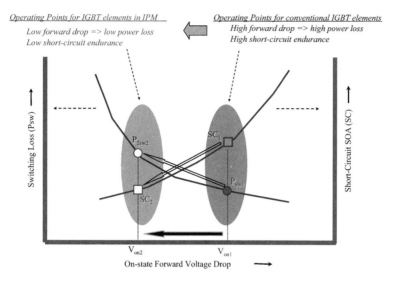

Figure 3.29 A graphic illustration showing the fundamental motive behind IPM invention. Reprinted from Ref. [83] with permission from the Institute of Electrical Engineers of Japan.

element–related problem within the IGBT device structure was the driver behind invention of the IPM's fundamental concept in the second half of the 1980s, as illustrated in Fig. 3.29.

In an IPM, protection of each internal IGBT element is done uniquely by monitoring the main device's current density condition through an integrated current-sensing scheme and feeding back the information to an integrated control integrated circuit (IC) (low-voltage ASIC). The diagnostic function inside the ASIC checks and performs the necessary overcurrent (short-circuit) protection under abnormal conditions independently and swiftly on a real-time basis. To contain the protected IGBT's voltage-current trace within a tolerable limit (SOA), the integrated protection scheme in an IPM is also designed to perform an overcurrent shutdown in an intelligent way. This is done either by reducing the protected element's gate voltage to a lower level or by activating a separate gate circuit to control the gate charge extraction process and to suppress the turn-off di/dt of the device's collector current and, thus, to avoid any destructive collector voltage overshoot.

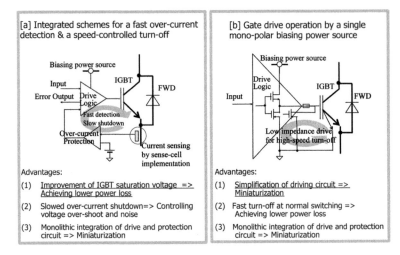

Figure 3.30 Details of the fundamental IPM functionalities implemented to realize the concept of "local monitoring and safe control of power transistor (IGBT) operation on a real-time basis." Reprinted from Ref. [83] with permission from the Institute of Electrical Engineers of Japan.

Figure 3.30 describes the fundamentals of an IPM's functional concept schematically. As explained before, the first feature (Fig. 3.30a) of an IPM created by its invention is that an integrated sensing and protection circuit scheme detects any overcurrent situation of an internal IGBT power switch almost instantaneously and turns off its current safely at a subdued shutdown speed to suppress a destructive surge voltage. The scheme in Fig. 3.30a thus effectively enlarges the SOA of the internal IGBT. The second feature is that only a monopolar positive supply (e.g., 15 V) is required to drive and protect the internal IGBTs of an IPM. This is achieved by creating a low-impedance circuit across the gate-emitter of each internal IGBT at its turn-off switching and during its turned-off state and thus eliminating the need for a negative power source. The feature in Fig. 3.30b thus contributes to the downsizing of the application system by simplifying the IGBT drive circuitry. Also, a fast turn-off type internal gate control scheme reduces the switching losses of the IPM and, thereby, effectively solves the other trade-off issue between on-state loss and turn-off switching loss in the case of conventional IGBT by appropriately reducing both.

Such an integration concept implemented in an IPM has proven that an application system's size can be drastically reduced, and, at the same time, its performance and reliability can be adequately improved by applying an appropriate IPM. The monolithic low-voltage application-specific integrated circuits (LVICs) used in IPMs were developed using large-scale-integration (LSI) process technologies. Later on high-voltage application-specific integrated circuits (HVICs) were also developed for further functionality enhancements and applied in several versions of low-power-category IPMs, together with optimized packaging techniques and circuit layout designs.

3.3.2.2 Chip technologies driven by the IPM evolution

3.3.2.2.1 *Power chip* MOS-gated device concepts have played a primary role in driving the growth of power semiconductors since the late 1980s, thanks to the revolutionary achievements made by IGBTs and IPMs through several generations of refinements [22–49, 53–80]. The IPMs continued to offer an edge over equivalent IGBT modules in terms of performance, reliability, and system compactness by virtue of their integrated intelligence [83]. In both cases of product evolution, refinement efforts were particularly aimed at improving performances of MOS-gated active switches related to reduction of on-state voltage/resistance drop, achievement of higher current conduction density, reduction of power losses for faster switching operation, and enhancement of the safe operation area to withstand hard short-circuit stresses.

A chronograph of the figure-of-merit (FOM) improvement for various generations of IGBT devices is shown in Fig. 3.31 normalized to that of the first generation. An inset at the top of the figure describes the key technologies that contributed to this progress in the past. Likewise, an inset at the bottom describes the new key technologies that will be effective for future improvement of the FOM.

Figure 3.32 shows some typical trade-off relations between normalized values of forward-voltage drop and turn-off switching energy per pulse of switching operation for different generations of IGBT elements developed for applications in the category of 600 V

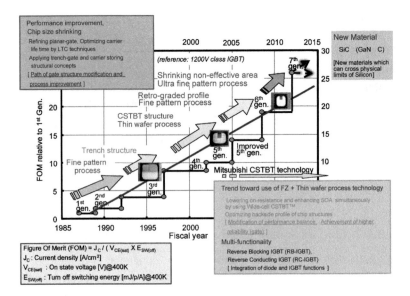

Figure 3.31 The chronograph of the IGBT's FOM improvement to date and the outlook on future possibilities. Reprinted from Ref. [83] with permission from the Institute of Electrical Engineers of Japan.

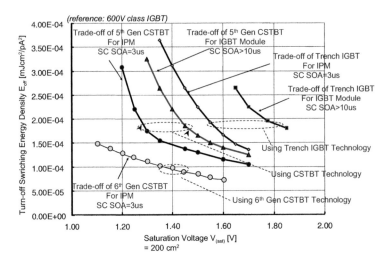

Figure 3.32 Some typical trade-off relations of the IGBT. Reprinted from Ref. [83] with permission from the Institute of Electrical Engineers of Japan.

class IGBT module equivalent and IPMs. As revealed, in the case of IGBT elements used in normal power modules, the trade-off curve of a particular generations lies at a higher level compared to that used in an equivalent IPM. This distinction in the trade-off relationship is directly related to the difference in short-circuit endurance between the two IGBT element designs, as discussed earlier. For the reasons described before, an IGBT element for an IPM application is not required to exhibit a high degree of short-circuit-withstanding capability compared to that required for a normal nonintelligent power module application. Thus, an IPM of any generation has exhibited much better power loss characteristic than a normal IGBT module of the same generation so far. Although the IGBT's performance has greatly improved in recent years and devices with a usable short-circuit performance and significantly low forward-voltage drop characteristic have been made available by the introduction of the latest versions, such as the Carrier Stored Trench-gate Bipolar Transistor (CSTBT^{TM}), the IPM concept still excels at improved performance of power modules as the trend toward integrating more functions and achieving higher power density in system designs continues.

3.3.2.2.2 HVIC One of the key technologies supporting the evolution of IPMs is considered to be the concept of integrating HVICs instead of the low-voltage ASICs discussed before. As highlighted in Section 3.3.1 and in Fig. 3.33, a typical HVIC architecture monolithically integrates all functions necessary to transmit signals from a controller at low potential to the gate of an IGBT, which can be either at the same potential or floating at a level several hundreds of volts higher than that.

Principally, as also explained in Chapter 3, such a high-voltage-level swinging operation to drive an IGBT gate on a power inverter bridge circuit is made possible by a pair of high-voltage-level shifters (creation of high-voltage small-signal type n-channel MOSFETs) and proper intra-circuit-island isolation structures embedded in the chip architecture. An HVIC can also perform reverse signal transmission from a similar high-potential level to a similar lower level of a controller by incorporating reverse-level shifters (high-voltage small-signal-type p-channel MOSFETs) in the same architecture.

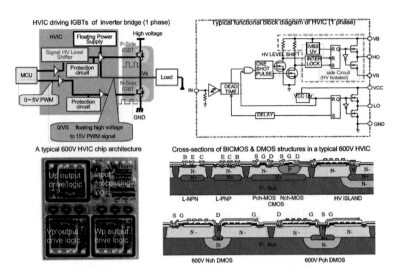

Figure 3.33 Architectural features of a typical HVIC and its function of driving high-voltage IGBTs on an inverter bridge circuit. Reprinted from Ref. [83] with permission from the Institute of Electrical Engineers of Japan.

Such unique signal transmission function, in fact, can eliminate the need for optical couplers or signal transformers used for the purpose of electrical signal isolation. Furthermore, virtually all kinds of logic functions, signal processing, and amplification functions, including direct gate driving capability for medium-range MOS-gated power devices can be integrated in a HVIC. Such capabilities of HVICs make them an ideal choice for integration in IPMs to create advanced power modules. Presently, HVIC processes of 600 V and 1200 V classes have been developed [68, 71–73, 83]. High-voltage ASICs generated by these processes are applied in the transfer-molded-package-type DIPIPM series rated up to 50 A and 1200 V. Some of the latest processes are capable of integrating sophisticated data storing functions, such as an erasable programmable read-only memory (EPROM) unit, and such progresses promise development of smarter and more compact highly reliable IPMs in the future. Figure 3.33 is an illustration showing the functional performance, chip architecture, and cell cross-sectional description of a typical 600 V class HVIC technology, which is based on a unique junction isolation structural concept.

3.3.2.2.3 *Functional integration technology*

Figure 3.34 shows the basic block diagram of an IPM, highlighting its concept and main advantages. Each IGBT chip integrated in an IPM typically contains an on-chip current sensor, which feeds back collector current information to an internal ASIC for processing and performing overcurrent and short-circuit protection functions on a real-time basis, as described earlier.

In the case of a normal IGBT module application, external current transformers are used to sense current and the sensed current value is fed back to a controller (central processing unit [CPU]) through an amplifier and A/D converter circuitry. The controller monitors current and performs signal shutdown operation to protect the related IGBT from overcurrent. The signal shutdown information is fed back to the related IGBT's gate via a gate signal isolation unit, and thus the whole process, from current sensing to IGBT current turn-off action, takes more than 10 μs typically. Therefore, the IGBT elements designed for nonintelligent power module applications are required to have more than 10 μs of short-circuit endurance and a sufficiently wide SOA characteristic for ensuring the system's reliability. In other words, the power loss performance of IGBT elements in such modules is sacrificed to some extent for reasons described earlier. In the case of an IPM, an overcurrent (e.g., short-circuit current) the status of any internal IGBT can be determined typically within 3 μs by the integrated protection scheme explained earlier and described in Fig. 3.34. This swiftness in protection operation and voltage overshoot suppression scheme at short-circuit shutdown eliminates the need for an internal IGBT element to possess any inherent short-circuit endurance. Therefore, the IGBT elements for IPM applications can be designed to have a very low forward-voltage drop characteristic, down almost to the ideal level that can be extracted from the device structure without caring about the internal parasitic thyristor.

Another typical function integrated in an IPM is the over-temperature protection scheme. Up to the third-generation stage of the IPM evolution, a thermistor, or a thermal sensor, was mounted inside on the base plate surface at a specific position. Thus an average base plate temperature (also termed as the module's case temperature, T_c, on the data sheets) used to be monitored. However,

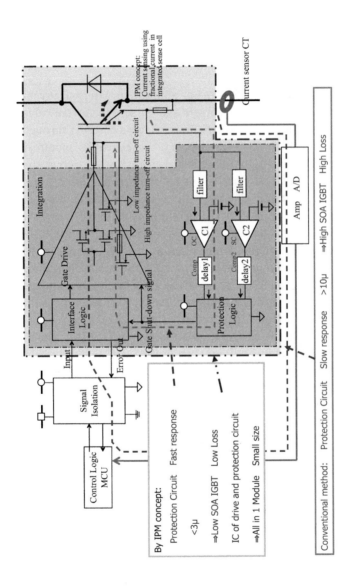

Figure 3.34 Basic block diagram of a typical IPM's internal circuitry featuring a comparison with conventional methods of driving and protection schemes for IGBTs. Reprinted from Ref. [83] with permission from the Institute of Electrical Engineers of Japan.

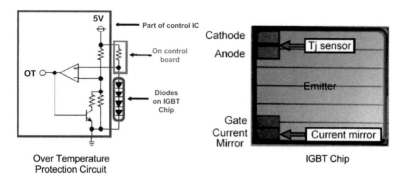

Figure 3.35 For further improvement of the IPM's performance and reliability, an on-chip diode sensor scheme was developed by Mitsubishi Electric and this added feature has been in use since the fifth-generation IPM product line-up. Reprinted from Ref. [83] with permission from the Institute of Electrical Engineers of Japan.

the heat dissipation time constant of base plates being very large, the thermistor sensing and temperature protection scheme was not appropriate for protecting against a localized or a rapidly rising temperature of certain internal power chip under some specific operation mode, such as a fast-overloading or a motor-lock situation, of the applied inverter system. To avoid this drawback and make the internal overtemperature protection function more efficient, an on-chip sensing scheme was developed and introduced in Mitsubishi's fifth-generation IPM series. Figure 3.35 schematically explains this protection scheme and shows an actual chip architecture depicting the location of such integrated sensing elements (a string of series-connected small-signal-type polysilicon diodes) on the chip surface.

Another example of IPM advancement achieved at the fifth-generation level in terms of intelligent functionality is integration of a unique feedback loop to control the IGBT's gate impedance depending on its current switching level. This new function was designed to change the switching speed of the driven IGBT from high to low, or vice versa, depending on its switching current level (low or high) to suppress electromagnetic interference (EMI) noise generation effectively without sacrificing much by increased switching losses. The main cause of noise radiated from the power module is a fast switching operation performed by the internal

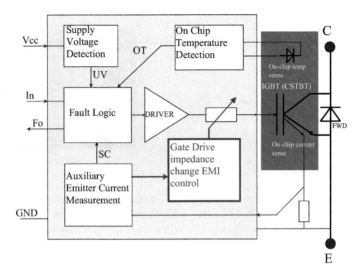

Figure 3.36 Internal block diagram of Mitsubishi's fifth-generation IPM having a feedback loop for adjusting the IGBT's gate impedance at its low-current-level switching for suppressing generated EMI noise effectively. Reprinted from Ref. [83] with permission from the Institute of Electrical Engineers of Japan.

IGBT power elements. Fast switching characteristics are desirable for reducing switching losses in power conversion applications. However, the noise level increases with the dv/dt generated due to the speed of the switching transients and, thus, the trade-off relationship between noise generation and switching losses has been a difficult design issue in power conversion system applications. The new feature in the latest fifth-generation IPM series solves this issue adequately by changing the gate switching speed at a preset value of the switched collector current level (Fig. 3.36).

Typically, the gate speed changing-over point is set at the 50% level of the rated collector current, I_{Cr}, of a module. An on-chip current sensor sends back the IGBT's current level of information to a new ASIC integrated inside each IPM. At 50% of the I_{Cr} level, the particular IGBT's gate driving speed is automatically changed from fast mode to slow mode, when the switched current goes below that level and when the opposite happens. The gate speed control method adopted here is based on gate charge controlling of the driven IGBT element. The fast gate-drive mode used for high-current switching reduces switching power loss, which is predominant at that current range. On the contrary, the slow gate-drive mode for low-current switching reduces the radiated noise level, which becomes more alarming than the switching loss increment at that current range. In actual applications, the fifth-generation IPM series has exhibited that the radiated noise can be reduced by about 8 dB on an average without increasing the total power losses over the full range of the assigned operating load current.

3.3.2.2.4 Package construction technology   Figure 3.37 shows constructional features of a typical medium-power-class six-pack IPM, which has been made commercially available since the mid-1990s. Through various generations of device evolution so far, the IPM family is able to cover up to 3.3 kV voltage class using this basic packaging concept. The highest ratings made available on a single device are 1.2 kA and 3.3 kV. Figure 3.38 shows the package structure of one of the more recent versions of the

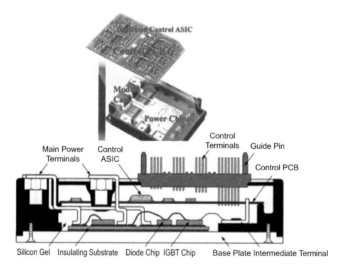

Figure 3.37 Typical package structure and constructional features of Mitsubishi's first-generation IPM series used for medium-power-range ratings (50–200 A/600–1200 V/three-phase output). Reprinted from Ref. [83] with permission from the Institute of Electrical Engineers of Japan.

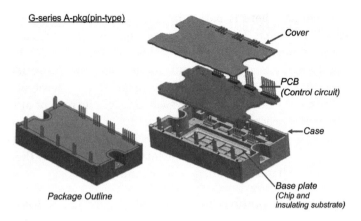

Figure 3.38 Typical package structure and constructional features of Mitsubishi's G-series IPMs (100–300 A/600–1200 V/three-phase output). Reprinted from Ref. [83] with permission from the Institute of Electrical Engineers of Japan.

IPM family, introduced by Mitsubishi in 2010s. This structure used fewer components and exhibited higher capabilities than its previous generations in terms of power cycling and thermal cycling endurances.

Also, continuous demand for further compactness (hence a lower system cost) and high-quality performance led to the development of a super-compact-size DIPIPM using transfer-molded packaging construction, targeted at low-power inverter applications, particularly those for household appliances [72, 73, 83]. Figure 3.39 shows an illustrative photograph of this DIPIPM with its structural features and functionalities.

In the following years, further miniaturized versions of this transfer-molded packaging concept, providing high-density silicon power solutions for up to 1200 V class using the latest power chip and HVIC technologies, were successfully developed and commercialized.

One of the major trends in the DIPIPM evolution is to increase the power-density-handling capability of the package. To achieve it the thermal impedance of the insulation structure becomes a critical issue. Through R&D efforts, a new thermal sheet has been created to replace the original epoxy resin insulation method used in the

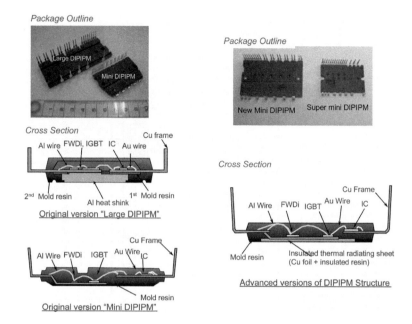

Figure 3.39 Features of Mitsubishi's transfer-molded DIPIPM series package for a low-end power range (5–50 A/600 V/three-phase output class). Reprinted from Ref. [83] with permission from the Institute of Electrical Engineers of Japan.

former generations of the device family. The sectional drawings shown on the right-hand side of Fig. 3.38 explain this feature. By applying the new insulation concept, the power-handling capability of the DIPIPM family increased by two times, opening the possibility for further improvement.

3.3.2.2.5 Advantages offered by the IPM Figure 3.40 summarizes some of the major advantages of applying an IPM compared to an IGBT-based system design. The comparison made here is based on a three-phase variable-voltage variable-frequency inverter system application. The difference in IGBT elements' FOM values leads to the difference in power loss and system size volume. Using an IPM the total system's power loss and size (volume) can both be reduced by about 20% and 10%, respectively. Along with this, a very reliable system can be built taking much shorter

228 | Applied Power Device Family

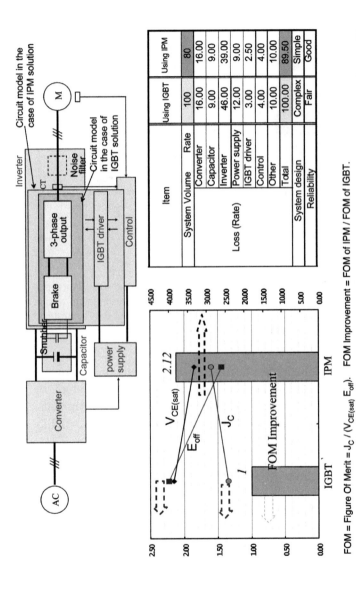

Figure 3.40 Merits of IPM application in an inverter system. Reprinted from Ref. [83] with permission from the Institute of Electrical Engineers of Japan.

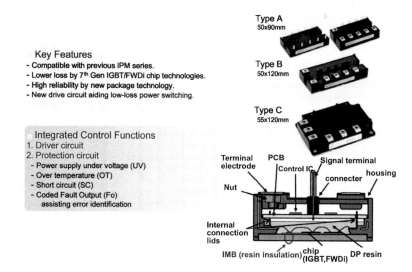

Figure 3.41 Features of an advanced G1 series IPM in mid-2010 (courtesy Mitsubishi Electric).

design or development time. Thus, IPMs, providing such important advantages to the application sides, will undoubtedly continue to remain as the key component solution for most power electronics applications.

3.3.2.2.6 IPMs of advanced generation In Fig. 3.41, features of an advanced IPM series introduced by Mitsubishi in the mid-2010s are described. The G1 family described here exhibits a high level of advancement that has been achieved by virtue of decades of technological innovations in power chips, peripheral circuitry, and IC designs and packaging structures, all based on the original platform of the IPM concept.

The power chip technology used in the G1 series IPM is of seventh-generation level. Each arm-switch of the three-phase power output internal circuit shown in Fig. 3.42 is configured by a pair of antiparallel connected seventh-generation CSTBT and FWD chips. The transistor and diode elements for the dynamic braking function shown on the extreme left of the internal circuit are also composed of the same generation CSTBT and freewheeling diode (FWD) chips. Each CSTBT chip has on-chip schemes for main current and chip

230 Applied Power Device Family

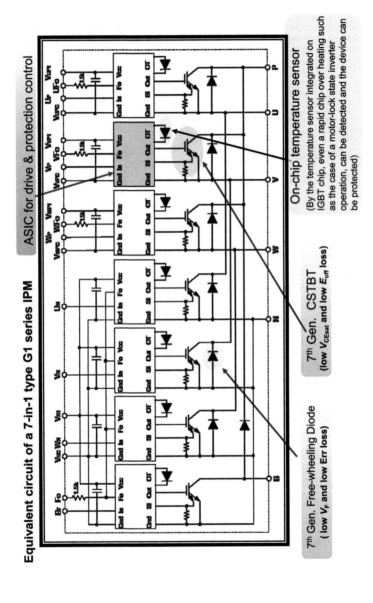

Figure 3.42 Description of the internal circuit block diagram and features of key integrated elements of a typical G1 series IPM (courtesy Mitsubishi Electric).

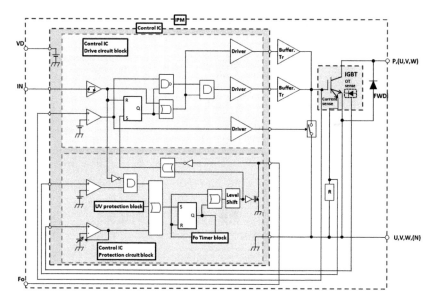

Figure 3.43 Circuit block diagram equivalent to one typical arm-switch in a G1 series advanced IPM (courtesy Mitsubishi Electric). The figure also highlights key control logic functions integrated in the ASIC.

temperature sensing. The core of each CSTBT's drive and protection circuitry is a dedicated ASIC, as shown in Fig. 3.42. Details of the ASIC's internal circuit configuration are shown in Fig. 3.43.

To drive a high-current CSTBT chip of 100–200 A/1200 V class arm-switch inside the module, additional buffers are used at the ASIC's output. The input interfaces use a Schemitt trigger to elevate the noise margin in the operation. With the buffer circuit addition option, the ASIC is thus applied as a common solution for a wide current rating range of the G1 IPM series.

Operation of various protection logics together with the CSTBT gate driving feature can be understood from the time charts given in Figs. 3.44 and 3.45. As revealed, the gate drive signal is shut off when either a power supply undervoltage fault state, a short-circuit fault state, or an overtemperature fault state occurs. The type of fault can be identified from the distinguishable duration of each fault output (FO), as shown in Fig. 3.45. This coded-error-type FO function makes troubleshooting on the system control side much simpler.

232 | Applied Power Device Family

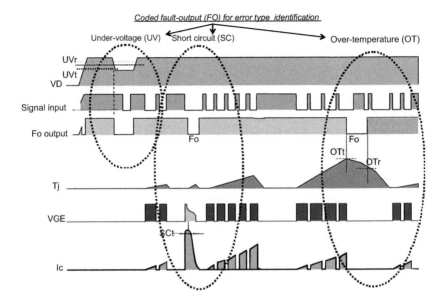

Figure 3.44 Operation time chart of protection functions integrated with the advanced series IPM (courtesy Mitsubishi Electric).

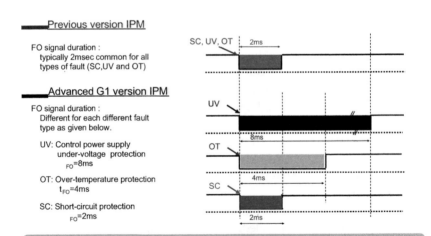

Figure 3.45 Fault output time chart comparison between the advanced G1 series IPM and the previous IPM series (courtesy Mitsubishi Electric).

State-of-the-Art Key Power Module Components | 233

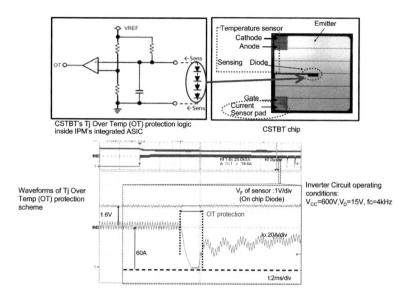

Figure 3.46 Overtemperature protection scheme used in the advanced G1 series IPM (courtesy Mitsubishi Electric).

The overtemperature protection scheme used in a G1 series IPM is shown in Fig. 3.46. The scheme is similar to the basic method used in the previous generations of IPMs, as explained earlier in this chapter (refer to Fig. 3.35). As shown in the figure, a string of series-connected diode sensors is embedded on-chip at the center of the IGBT (CSTBT) power chip. The variation of the forward-voltage drop v_f of the series-connected diode string scales to the temperature variation of the power transistor. The v_f signal is processed through a comparator having a reference voltage, which is set to be equal to the IGBT chip overtemperature trip level, inside the ASIC. The output of the comparator is further processed through the logic circuit integrated inside ASIC to execute the IGBT gate drive signal shutdown and error output, depending on the operation status. This shutdown operation can be understood from the waveforms inserted in Fig. 3.46 as an example.

The current-monitoring and short-circuit protection scheme used in the G1 series IPM is shown in Fig. 3.47. The scheme is similar to the basic method used in the previous IPMs explained earlier in this chapter (refer to Fig. 3.35). As mentioned before, current

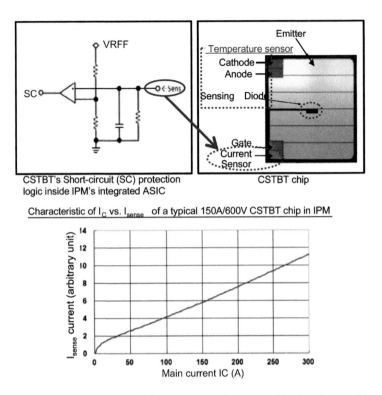

Figure 3.47 Short-circuit (SC) protection scheme used in the advanced G1 series IPM (courtesy Mitsubishi Electric).

monitoring is done by diverting a fraction of the total emitter current of an IGBT power chip through one or a few cell emitters, which are isolated from the rest of the chip cells. The segregated fractional current has a linear relationship to the main IGBT current, as shown in the graph inserted in Fig. 3.47, and forms a voltage drop across an external current-sense register, which automatically becomes the sensed signal proportional to the main current. The sensed voltage is processed through a comparator having a reference voltage, which is set to be equal to the IGBT chip short-circuit trip level, inside the ASIC. The output of the comparator is further processed through the logic circuit integrated inside the ASIC to execute a quick IGBT gate drive signal shutdown and provide an error output, depending on the operation status. The gate shutdown process is, however, done by adopting a slow turn-off driving-off method to suppress surge overvoltage across the protected IGBT similar to that of the previous IPMs. This shutdown operation can be understood from the time chart in Fig. 3.45 as an example.

Achieving high accuracy and appropriate linearity in sensing over a wide range of both temperature and current variations and also across all production lots for both overtemperature and short-circuit protection schemes is the important design aspect of the scheme that complicates because of properties of different materials and components involved.

3.3.2.2.7 A short projection of future trend related to functionality integration Other IPM target applications triggered by the energy concerns are in the field of transportation locomotives and electric vehicles, hybrid electric vehicles, fuel cell electric vehicles, etc. [74–78]. For this rapidly increasing range of applications, very high-voltage intelligent power modules (HV-IPMs), electric vehicle intelligent power modules (EV-IPMs), and hybrid electric vehicle intelligent power modules (HEV-IPMs) have been also launched since the late 1990s and have demonstrated their capability of meeting the severe reliability requirements demanded by these applications while at the same time realizing compact and smart system designs.

In the future, the concepts of power integration extending over the basic IPM-like platform are expected to be followed even more widely, covering a very broad portion in the power-handling capability and integrated functionality plane as shown in Fig. 3.48. From this figure one can see that, depending on the target application, the functionality of the IPM ranges from a simple but intelligent switch level to a complete system level having its own controller and self-biasing supplies built inside. As the technology trend continues to achieve new breakthroughs in the areas of silicon processing, realizing cost-effective manufacturing of new semiconductor materials such as silicon carbide (SiC), packaging materials, assembly techniques, system simulation, and circuit isolation techniques, the power semiconductor devices are expected to grow further primarily on the basis of the IPM-like system integration concept.

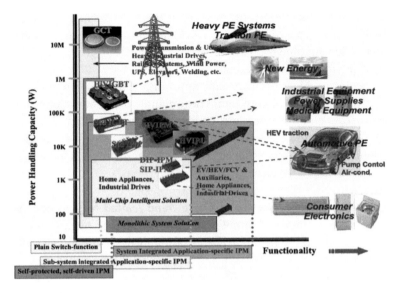

Figure 3.48   Versatility and expanding applications of power devices.

The topic of IPM evolution and functionality integration is further taken up in Chapter 4.

3.4 Tips on and Guidelines of Applying Power Modules

This section deals with the application aspects of major categories of power modules, providing some tips and guidelines for designing them in power electronics equipment effectively. The first part takes up common design issues involved in using high-switching-speed power module devices, such as IGBT modules and IPMs, in an inverter bridge circuit configuration.

3.4.1 Formulas Common for Power Device Driver Circuit Designs

The following formulas are often used in the hardware design of the power device drive circuit. Although we have the differential equation, taken out from only a transient change in the inverter of the analysis, and $V = L \times dI/dt$, $I = C \times dV/dt$ in the case of a linear region (linear approximation), four arithmetic operations can be used. There are many cases where it is valid to consider the range.

Ohm's law for the resistive element in the circuit is $V = I \times R$, where V is the circuit voltage, I is the circuit current, and R is the circuit resistance.

The equation for the inductive element in the circuit is $v = L \times di/dt$, where L is the circuit inductance.

The equation for the capacitive element in the circuit is $i = C \times dv/dt$, where C is the circuit capacitance.

3.4.2 Structure and Operation of the IGBT Device Used in a Power Module

In the previous chapters, details of the IGBT device structure and the physics of its operation principle are presented. In this section, the device's main structural and operation features are briefly reviewed for a quicker understanding before starting on application design aspects. From the preceding chapters it is clear that the IGBT is a switching transistor that is controlled by voltage applied to the gate terminal. The device operation and structure are similar to those of an insulated gate field-effect transistor, more commonly known as a MOSFET. In other words, a typical IGBT device structure is one born from that of a typical vertical double-diffused metal-oxide-semiconductor field-effect transistor (VD-MOSFET). The principal difference between the two device types is that the IGBT uses conductivity modulation to reduce on-state conduction losses. As in the cell structure examples shown in Fig. 3.49, n-channel-type IGBTs are those formed by essentially replacing the n^+ drain of the n-channel power VD-MOSFET with the p^+ collector layer. Even though the two devices have structural similarities, their operation principles are quite different, particularly when it relates to their turn-off switching operation on an inductive load inverter bridge circuit.

Figure 3.49 also shows equivalent circuits representing the two device structures, an IGBT cell and a power MOSFET cell, including

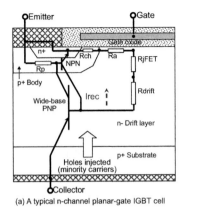 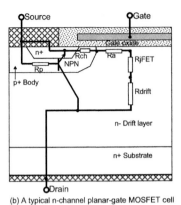

Figure 3.49 Structures and equivalent circuits of IGBT and power MOSFET cells indicating parasitic elements.

the parasitic element involved. Because the MOSFET and the IGBT are voltage-controlled devices, they require a voltage only on the gate to maintain conduction through the device. The IGBT has one junction more than the MOSFET, and this junction allows higher blocking voltage and conductivity modulation, as described later, during conduction. This additional junction in the IGBT does limit switching frequency, however.

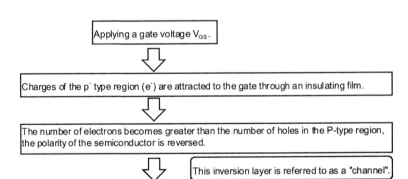

Figure 3.50 Basic operation of a power MOSFET (in the case of the n-channel).

Tips on and Guidelines for Applying Power Modules | 239

| Current flows through the channel of the MOSFET. |

| Because of the channel resistance, voltage of the channel beneath the gate of the drain side is increased. |

| The channel length reduces by the voltage effect directly under the gate. |

| When the channel current is increased, finally channel collapses. For this reason, the channel is not allowed to conduct more than a certain value of the current. |

| Increasing the gate-source voltage (V_{GS}), the channel immediately below the gate spreads its voltage difference with the gate and, thereby, makes room for a larger value of constant current (I_D) to flow through the channel. |

          | Increasing the V_{DS}, I_D also increases. |

| Furthermore, increasing the gate-source voltage (V_{GS}), the constant current value becomes larger than the load current and the device gets in to an on-state at a certain on-resistance. |

| At on-state the device shows a resistance characteristic:
$V_{DS} = I_D \cdot r_{DS(ON)}$
where, $r_{DS(ON)}$ is termed as the on-resistance |

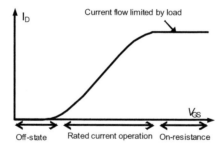

Figure 3.51 Power MOSFET operation at a rated current (in the case of the n-channel).

240 | *Applied Power Device Family*

When the device is on, the collector is at a higher voltage than the emitter and therefore minority carriers are injected from the collector p^+-region into the collector bulk region (n^+-buffer layer and collector n-region). The charges reduce the collector bulk region resistance and thus collector to emitter voltage drop is reduced (relative to V_{DS} (on) of the MOSFET). When a positive gate voltage is first applied, a gate current flows until the gate capacitance is charged and the gate voltage rises to the on level. When the gate voltage is removed, the charges injected into the collector bulk region must be removed before a high voltage can be blocked. The operation mechanisms of the two insulated gate devices are reviewed in the following section.

3.4.2.1 Review of power MOSFET and IGBT basic operation principle

In this section a quick review of the two insulated gate transistor devices is made using charts and illustrations to help understand their application aspects.

Note: It is possible to control the drain current by varying the gate voltage applied with respect to the source. Furthermore, when a negative gate-source voltage or a voltage below a threshold value needed for a channel inversion mechanism is applied, holes in the p-region below the gate structure are attracted, causing a reversal of the channel to the original p-type and the drain current ceases to flow and the device turns off.

A power MOSFET typically operates by charging the input capacitances (C_{GS}, C_{GD}) through the gate resistance R_G by applying a voltage V_{GS} voltage across the gate-source terminals to feed charges for the process. Switching speed transients (rise and fall times for the main drain current at turn-on and turn-off operations, respectively) by changing the gate control voltage V_{GS} depends on charging time required for the capacitances.

As shown in Fig. 3.53, C_{GD} shows dependency on the drain-source bias voltage V_{DS}. Therefore, C_{GD} dynamically influences each on and off switching transient of every switching cycle, where V_{DS} goes high to low and low to high, respectively. During this C_{GD} charging, V_{GS} stays at a constant value, which is typically slightly higher than the

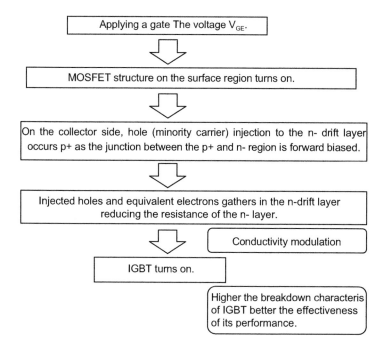

Figure 3.52 IGBT basic operation principle (in the case of the n-channel).

device's gate-source threshold voltage, $V_{GS(th)}$, for a specific time, allowing completion of the charging process. This phenomenon is known as the Miller effect, and it can be observed on a typical V_{GS} transient waveform captured at a turn-on or a turn-off switching operation. The extra delay time that appears at a turn-on or turn-off cycle is primarily effected by the charging of C_{GS}, which does not vary so much with V_{DS} as reveals from the capacitances characteristic in Fig. 3.53.

As a thumb rule, the driving power to be supplied from the gate drive circuit can be calculated as follows:

- Average current $I_G(\text{typ}) = k \times Q_G \times f_C$ where, $k = 1.3$ is typical.
- Peak current $I_{Gpeak} = V_{GS}/(R_G + \text{internal impedance})$.
- Drive power $W(\text{typ}) = V_{GS} \times Q_G \times f_C$ where Q_G is the gate charge value obtained from characteristics shown before and f_C is the switching frequency of the design.

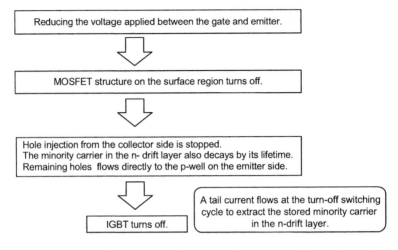

Figure 3.53 IGBT basic turn-off operation (in the case of the n-channel).

As a power MOSFET is a charge-controlled device, its switching speed can be changed by V_{GS} biasing condition and the gate circuit impedance (e.g., R_G) design made on the driver circuit. For achieving a fast switching speed design, R_G is normally chosen to be very low. For example, if the 7 A/900 V device characterized in the example is applied to a resistive load switching circuit having supply voltage $V_{DD} = 200$ V, load current $I_L = 9$ A, gate drive voltage $V_{GS} = 0$ V to +10 V at turn-on and +10 V to 0 V at turn-off, and gate resistance $R_G = 25$ Ω originally, the switching speed can be made three times faster by changing gate resistance to $R_G = 5$ Ω. However, when choosing a low R_G, you must be cautious about unwanted oscillations appearing on voltage and current waveforms due to parasitic inductances and capacitances in the gate and main circuits.

The IGBT is also a gate-charge-controlled voltage-driven device like a power MOSFET. It also has terminal capacitances due to its insulated gate structural feature, and it exhibits a MOSFET-like gate charge and capacitances characteristics, as shown in the example earlier. To use an IGBT in a switching circuit, the gate drive circuit should have power to charge and discharge its input capacitance C_{ies} instantly from the gate side at each turn-on and turn-off cycle. For a high-switching-frequency operation, the applied IGBT's gate-driver output power capability should be checked as the required

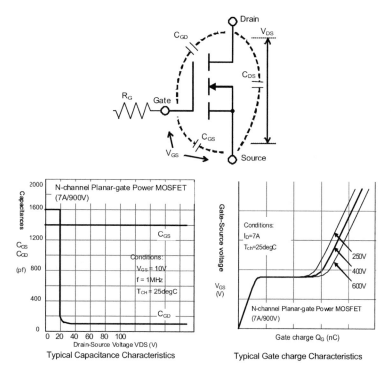

Figure 3.54 Capacitance/gate-charge characteristics of the power MOSFET.

average driving power in such a case increases. The gate-emitter capacitance component C_{GE} accounts for a large proportion of the equivalent input capacitance value. This capacitance component is created mainly by the gate-oxide layer formed by SiO_2 like a power MOSFET. A pure insulating gate-oxide layer, like the SiO_2 film, does conduct any DC electricity, and no charging or discharging takes place through the gate circuit when the device is either in a fully on state or in a fully off state. When alternating current (AC) or rapidly varying transient electricity (e.g., gate voltage ramp-up/ramp-down at turn-on/turn-off switching cycles) appears across the oxide capacitance, charges are transferred with an average power that varies in proportion to the frequency ($f = jwc$) of the change and with an instantaneous peak charging current in proportion to rate of rise or fall of the voltage (dv/dt) associated

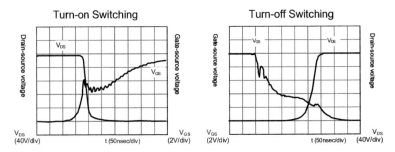

Figure 3.55 Typical waveforms of a 7 A/900 V n-channel planar-gate power MOSFET switching on a resistive load circuit.

with the transient change ($i = C dv/dt$). Like a power MOSFET switching operation, an IGBT also exhibits a Miller effect during its turn-on/turn-off switching, where the gate voltage stays constant dictated by C_{res} charging/discharging and a constant voltage plateau period appears as the gate voltage level stays constant. For designing the gate drive circuit the gate charge graph shown earlier is used. As a thumb rule, the driving power to be supplied from the gate drive circuit can be calculated as follows:

- The average current $I_{G(typ)} = k \times Q_G \times f_C$ where, $k = 1.3$ is typical.
- The peak current $I_{Gpeak} = V_{GE}/(R_G + \text{internal impedance})$.
- The drive power $W_{(typ)} = V_{GE} \times Q_G \times f_C$, where Q_G is the gate charge value obtained from characteristics shown before and f_C is the switching frequency of the design.

As the IGBT is also a charge-controlled device like a power MOSFET, its switching speed, especially the turn-on operation, can be changed by a V_{GE} bias (normally set at $+15$ V/-15 V with a 10% variation allowed on each level) and gate circuit impedance (e.g., R_G) design of the driver circuit. For achieving a fast turn-on switching speed, R_G can be chosen to be low. However, when choosing a low R_G, you need to be cautious about unwanted oscillations appearing on voltage and current waveforms due to parasitic inductances and capacitances in the gate and main circuits.

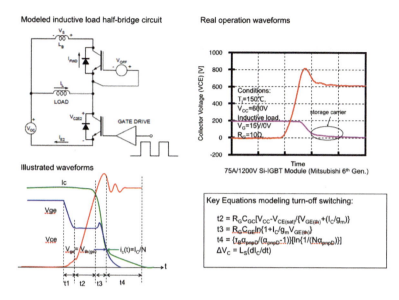

Figure 3.56 IGBT module turn-off switching on a half-bridge circuit with a parasitic bus inductance.

The surface emitter pattern of an IGBT chip is typically striped geometrically, in contrast to a typical power MOSFET's cell-based geometry. The IGBT uses the same small feature size advantages of the MOSFET, but the striped geometry offers a better turn-off switching performance and more ruggedness and immunity from the latch-up of the parasitic thyristor formed by the parasitic npn and pnp bipolar transistors shown in Fig. 3.49a. For a better understanding of the IGBT's internal capacitances and their roles in charge-driven dynamic characteristics and particularly the turn-off switching function of the device, which is basically a two-carrier-operated transistor having an excessive amount of minority carriers accumulated in its drift layer during an on-state and prior to a turn-off action, Figs. 3.56 and 3.57 are prepared. Key equations to model the turn-off switching operation of the device are given in these illustrations. The main operational behavior shown in each segmented period of an inductive load turn-off switching process of an IGBT is explained later, linking the key equations in Figs. 3.56 and 3.57.

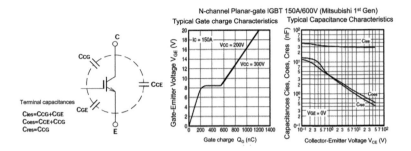

Figure 3.57 Capacitances/gate charge characteristics of an IGBT.

Period t₁: It is the transient interval when the terminal capacitances of the device, C_{GE} and C_{GC}, discharge and the gate-emitter-driving voltage, V_{GE}, falls to nearly to the device's gate threshold voltage, $V_{GE}(th)$ level.

Period t₂: It is the transient interval when the collector-emitter voltage, V_{CE} builds up, with V_{GE} staying at the device's gate-emitter threshold voltage due to the Miller effect explained before and causes the device channel to pinch off. The gate-emitter voltage V_{GE} and the gate charging current I_G can be expressed as given below.

$$V_{GE} = R_G C_{GC} \frac{dV_{CE}}{dt} = V_{GE}(th) + \frac{I_C}{g_m} \quad (3.3)$$

$$I_G = \frac{V_{GE}}{R_G} = \frac{V_{GE}(th) + I_C/g_m}{R_G} \quad (3.4)$$

g_m : trans-conductance.

Also, since V_{CE} rises linearly from the $V_{CE(sat)}$ level, aiding reverse charging of C_{GC} (i.e., $dV_{CE}/dt = I_G/C_{GC}$), the **t2** time can be expressed as given below.

$$t_2 = R_G C_{GC} \left(V_{CC} - \frac{V_{CE}(sat)}{V_{GE}(th) + I_C/g_m} \right) \quad (3.5)$$

Period t₃+t₄: These are the transient intervals during which V_{GE} continues to decay, dragging out internal gate charges, and V_{CE} surges up to a peak value above the source voltage V_{CC} due to the stray inductance L_S (sum of total inductance in the link between the source to the device). The V_{CE} surge peak can be given be estimated as follows:

$$\Delta V_C = L_S \frac{dI_C}{dt} \qquad (3.6)$$

$$I_C : \text{the collector current.}$$

During this action, if the energy stored in the L_S, which is $E_{ls} = L_S I_C^2/2$, is large, then the V_{CE} peak gets sustained due to pinch-off occurring in its drift layer causing an avalanche of carriers, or carrier-multiplication, feeding current to absorb the inductive energy. The V_{GE} decay and the electron current that flows through the MOS channel before its shut-off can be expressed as given next.

$$V_{GE} = \left(V_{GE}(th) + \frac{I_C}{g_m} \right) \exp \left(\frac{-t}{R_G C_{GE}} \right) \qquad (3.7)$$

$$I_{MOS} = (g_m V_{GE}(th) + I_C) \exp \left(\frac{-t}{R_G C_{GE}} \right) - g_m V_{GE}(th) \qquad (3.8)$$

Also, t_3 can be expressed as given next.

$$t_3 = R_G C_{GE} \ln \left(1 + \frac{I_C}{g_m V_{GE}(th)} \right) \qquad (3.9)$$

Furthermore, as I_{MOS} reduces gradually to zero by channel shut-off in the t_4 interval, the residual minority carriers follow a separate decaying process from there by a recombination mechanism with the pnp parasitic transistor in an open-base mode. This carrier decaying constitutes a tail current in the IGBT's turn-off current and is mainly due to the remaining charge Q in the base region (IGBT drift region) of the open-base pnp transistor, having a dynamic gain value defined as α_{pnpD} and a base charge transportation time defined as τ_B. The tail current of I_C can be expressed as given next.

$$I_C(t) = I_C(0) \exp \left(\frac{-t}{\alpha_{pnpD} \tau_B} \right) (t) \qquad (3.10)$$

The base charge transportation time is proportional to the lifetime τ_H of the minority carriers, which can be further expressed as follows:

$$\tau_B = k(\tau_H) \qquad (3.11)$$

Now, if the tail current is considered to be the collector current of the IGBT device to fall to a $1/N$ level of the current's starting value at the beginning of the turn-off process, then the t_4 time interval can be further expressed, as given next.

$$t_4 = \frac{\tau_B \alpha_{pnpD}}{\alpha_{pnpD} - 1} \ln \frac{1}{N \alpha_{pnpD}} \tag{3.12}$$

The IGBT's turn-off switching time thus gets affected by the tail current mechanism due to the inherent minority carrier extraction time explained earlier. To make an IGBT more rugged against latch-up failure by a parasitic thyristor's latch-up at turn-off or to design it to be of a fast turn-off switching type to fit the requirement of high-frequency switching applications, its structural design and process concept aimed at shortening t_3 and t_4 are important.

Considering the latch-up issue related to the IGBT's parasitic thyristor the following chart will provide explanatory details for a better understanding of the subject.

3.4.2.2 Review of a parasitic thyristor's destructive latch-up in an IGBT cell

Figures 3.58 and 3.59 are prepared to explain the existence of the parasitic components forming a thyristor that can latch up during the IGBT operation, causing a destructive phenomenon. Figure 3.58 reviews the behavior under a static on-state operation of an IGBT structure.

3.4.3 Power Circuit Design

In high-power systems, rapid turn-on and turn-off operations produce harsh dynamic conditions. The power circuit, snubbers, and gate drive must be designed to deal with extreme di/dt and dv/dt stresses. Excessive transient voltages can occur if the leakage inductance in the power circuit and snubbers is not minimized. Ground loops and capacitive coupling can cause serious noise problems. An appropriate mechanical and electrical layout is essential for reliable and efficient operation of IGBTs and IPMs.

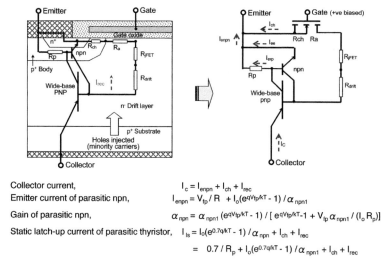

Figure 3.58 Equivalent circuit of an IGBT cell and condition for static latch-up.

Collector current, $I_c = I_{enpn} + I_{ch} + I_{rec}$
Emitter current of parasitic npn, $I_{enpn} = V_{fp}/R + I_0(e^{qV_{fp}/kT} - 1)/\alpha_{npn1}$
Gain of parasitic npn, $\alpha_{npn} = \alpha_{npn1}(e^{qV_{fp}/kT} - 1)/[e^{qV_{fp}/kT} - 1 + V_{fp}\alpha_{npn1}/(I_0 R_p)]$
Static latch-up current of parasitic thyristor, $I_{ls} = I_0(e^{0.7q/kT} - 1)/\alpha_{npn} + I_{ch} + I_{rec}$
$= 0.7/R_p + I_0(e^{0.7q/kT} - 1)/\alpha_{npn1} + I_{ch} + I_{rec}$

3.4.3.1 Turn-off surge voltage

The turn-off surge voltage is the transient voltage that occurs when the current through the IGBT is interrupted at turn-off. To examine this, consider the inductive load half-bridge circuit shown in Fig. 3.56. In this test circuit the top IGBT is biased off and the bottom device is switched on and off with a burst of pulses. Each time the lower device is turned on, the current in the inductive load (I_L) will increase. When the lower device is turned off, the current in the inductive load cannot change instantly. It must circulate through the FWD of the upper device. When the lower device turns back on, the load current will commutate back to the lower device and begin to ramp up again. If the circuit was ideal and had no parasitic inductance, the voltage across the lower device (V_{C2E2}) at turn-off would increase until it reached one diode drop above the bus voltage (V_{CC}). The upper device's FWD would then turn on, stopping the voltage from increasing further. Unfortunately, real power circuits have parasitic leakage inductances. In Fig. 3.56 a lump inductance (L_B) has been added to the half-bridge circuit to simulate the effect of a parasitic bus inductance. When the lower device turns off the

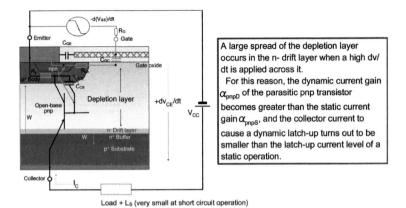

Figure 3.59 IGBT cell with an equivalent circuit explaining dynamic latch-up.

inductance L_B resists the commutation of the load current to the FWD of the upper device.

A voltage (V_S) equal to $L_B \times di/dt$ appears across L_B in opposition to increasing current in the bus. The polarity of this voltage is such that it adds to the DC bus voltage and appears across the lower IGBT as a surge voltage. In extreme cases, the surge voltage can exceed the IGBT's V_{CES} rating and cause it to fail. In a real application the parasitic inductance (L_S) is distributed throughout the power circuit, but the effect is the same.

3.4.3.2 Freewheeling diode recovery surge

A surge voltage similar to the turn-off surge can occur when the FWD recovers. Assume that the lower IGBT in Fig. 3.56 is off and that the load current (I_L) is circulating through the FWD of the upper IGBT. When the lower device turns on, the current in the FWD of the upper device (I_{FWD}) decreases as the load current begins to commutate to the lower device and becomes negative during reverse recovery of the FWD. When the FWD recovers, the current in the bus is quickly decreased to zero. The situation is similar to the turn-off operation

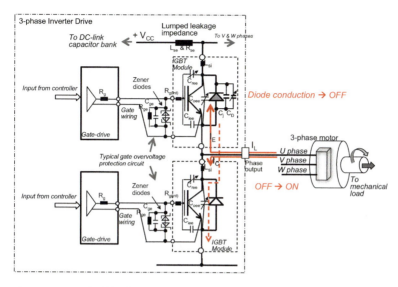

Figure 3.60 A half-bridge circuit modeled using one phase of a three-phase motor drive system to explain the freewheeling diode recovery process and its typical behavior under inductive load operation explained in Fig. 3.61.

described in Section 3.4.3.1. The parasitic lumped bus inductance (L_S) develops a surge voltage equal to $L_S \times di/dt$ in opposition to the decreasing current. In this case, the di/dt is related to the recovery characteristic of the FWD. Some fast-recovery diodes can develop extremely high recovery di/dt when they are hard recovered by the rapid turn-on of the lower IGBT. This condition, commonly referred to as "snappy" recovery, can cause very high transient voltages. In the earlier chapters of this book, the fundamental physics of a diode operation and its structural aspects are explained in detail. Here we will examine its reverse recovery operation considering a practical application.

Figures 3.60 and 3.61 are provided for a better understanding of an FWD's typical recovery process under a hard switching inductive load circuit operation (e.g., a phase circuit of a three-phase inverter drive system or a inductive load drive chopper control system). In this model, the FWD in the upper arm-switch of the U-phase bridge circuit is assumed to be completing its freewheeling action by forward-conducting the motor regeneration power either to the

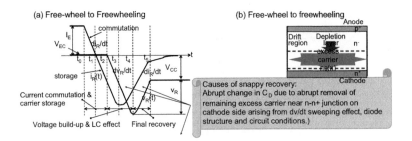

Figure 3.61 Freewheeling diode's reverse recovery process and general behavior.

DC link capacitor bank (not shown in the figure) and/or to the front-end AC side (also not shown in the figure) of the inverter system or by circulating current through other phases of the three-phase topology. The forward conduction of the diode completes when the lower IGBT is turned on by its gate drive signal. At that instant, the reverse phase current tends to flow through the turning on of the lower IGBT by a commutation process in which the upper-arm FWD ceases to conduct by turning through a reverse recovery process.

During the diode's reverse recovery, the arm circuit remains under a state shorted across the DC link until the full link voltage appears across the diode, swinging from the off-to-on IGBT of the lower arm. Such process involving inductive power commutation by a hard-switching operation requires the diode to be rugged and of the fast switching type, and its reverse recovery characteristic plays a very important role in the system design as it involves concerns related to trade-offs among switching power losses, steady-state power losses, voltage surge, oscillation in circuit current and voltages, and generation of EMI noises. Therefore, tuning of the diode structure and its fabrication process, together with module integration issues, are complicated and involve optimization through adequate research. In Fig. 3.61, details are given for understanding the behavioral peculiarities a diode exhibits during its reverse recovery turn-off cycle to remove the excess minority carriers stored in its structure. The various capacitive effects of the diode's internal capacitances and its interaction with peripheral circuit components and turn-on driving conditions for the bridge circuit opposite-arm IGBT are also explained in Fig. 3.61, providing details of the cause and effect of the diode's snappy behavior.

3.4.3.3 Design issues related to ground loops

Ground loops are caused when the gate drive or control signals share a return current path with the main current. During switching, the voltage is induced in power circuit leakage inductance by the high di/dt of the main current. When this happens, points in the circuit that should be at ground potential may, in fact, be several volts above ground. This voltage can appear on the gates of devices that are supposed to be biased off, causing them to turn on. To avoid this problem, careful referencing of gate drive and control circuits is required. In applications using large IGBT modules high di/dts make it increasingly difficult to avoid ground loop problems. Figure 3.62 shows the recommended circuit for IGBT modules rated 300 A or more. In this circuit separate isolated power supplies are

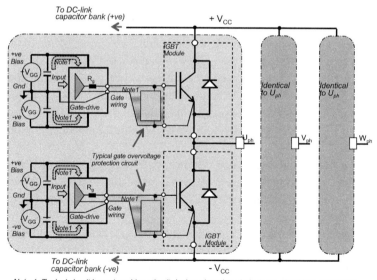

Note 1: Typical circuit loops in a driver circuit design where loop inductance should be minimized.

Figure 3.62 Driver circuit ground-loop design consideration for a three-phase inverter using a high-current-rated IGBT module (up to 300 A).

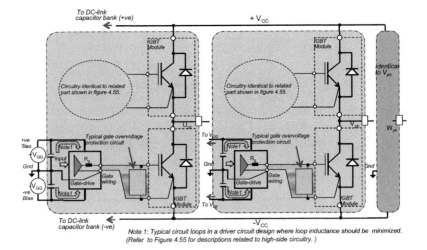

Figure 3.63 Driver circuit ground-loop design consideration for a three-phase inverter using medium- to high-current-rated IGBT module (up to 200 A).

used for each low side gate driver in order to eliminate ground loop problems. The figure also indicates the major circuit loops around the driver circuit of each IGBT module where extra care needs to be given to minimize stray inductances. In that sense, the driver should principally be laid close to the gate output terminals and connected directly. In the case a wire connection for a section of the circuit becomes an unavoidable choice, at least twisting of positive (+) and negative (−) wires should be adopted in the design to reduce loop inductance.

The figure also illustrates a typical gate-overvoltage protection scheme comprising back-to-back series connected zener diodes and a gate-emitter parallel resistor capacitor (RC) filter having a high impedance.

Figure 3.63 shows a circuit with potential ground loop problems. In this circuit the ground return for the gate drive passes through the main power bus. This circuit is suitable for use with low-current six-pack devices because they have minimal inductance in the negative bus and a relatively low-power circuit di/dt. However, even in this case a strong off-bias of −5 to −15 V is recommended. At higher operating currents, voltages induced in the bus during switching

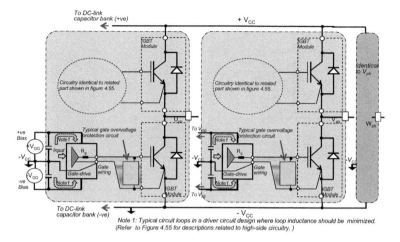

Figure 3.64 Driver circuit ground-loop design consideration for a three-phase inverter using a low-current six-pack-type IGBT module.

are likely to cause ground loop noise problems in the circuit of Fig. 3.63. Figure 3.64 shows the recommended connection of low side drivers using a single-gate-drive power supply. In this circuit, the ground loop noise is minimized through the use of auxiliary emitters and local power supply decoupling capacitors. This circuit is usually suitable for use with six-pack- or seven-pack-type three-phase-output-type modules rated up to about 200 A.

Note that in the case of any IPM product commercialized so far by Mitsubishi, a negative biasing of its gate circuit, neither for its turn-off operation nor for maintaining its turned-off state, is not necessary due to an appropriate impedance matching inherited in each device's internal driving scheme. However, the ground noise issue can still be significant if a component like a shunt resistor for inverter current monitoring/control or an overcurrent protection scheme is added in the N-bus ($-V_{CC}$) link of the inverter without an inappropriate layout design, increasing the stray inductance. The problem and solution can be explained further by taking the case of using a typical low-power 1.5 kW/200 V AC/three-phase inverter layout design using a 20 A/600 V six-pack IPM (e.g., a DIPIPM). Figure 3.65 illustrates the ground pattern layouts between the control circuit and the power circuit with an N-bus shunt resistor

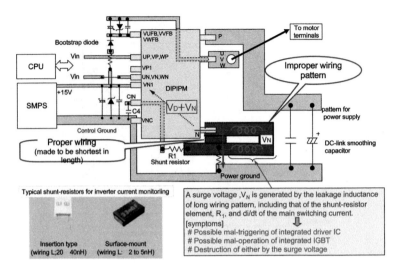

Figure 3.65 Effect of pattern design related to the N-terminal of a low-power inverter using a six-pack-type module (e.g., a DIPIPM) and a common N-bus of the system.

that can affect operation by inflicting ground noise (surge) voltages, which are compared by the waveforms of the actual operation in Fig. 3.66.

3.4.3.4 Reducing gate circuit inductance and avoiding mal-triggering

Wiring from the gate drive circuit output to power devices, the area that produces the closed circuit shown in Fig. 3.67 below the shaded portion, must implement a pattern designed as small as possible.

The generated voltage that causes mal-triggering by forming a mal-signal is simply related to the rate of change in the current (di/dt) circulating in the circuit encompassing the patterned area and the leakage inductance (L) in the whole loop or the change in the associated magnetic flux (ϕ) generation, which is in turn proportional to the same area. In other words, the relation can be shown as given next.

$$V = L\frac{di}{dt} = \frac{\phi}{dt} \qquad (3.13)$$

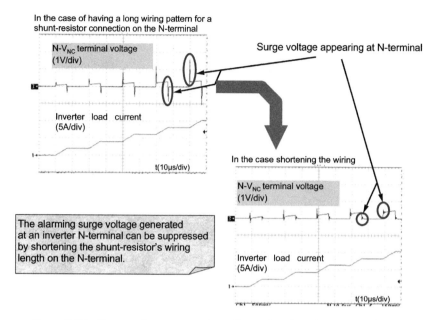

Figure 3.66 Surge voltage generation effect related to the pattern design explained in Fig. 3.65.

To design a driver circuit for high-speed switching, due attention to the leakage inductance, in particular, and the associated magnetic flux generation is required to be paid. In the case of using wiring connections from a driver's output terminals to an IGBT's gate terminal, the wiring length/size and their looping area become primary concerns. With long gate-circuit wires, the total leakage inductance (L) proportional to each wire's length/cross section, as well as its mutual inductance part related to the circuit loop and the generated ϕ, will increase. Reduction of leakage inductance becomes essential by way of pattern layout design to achieve driver operation less susceptible to the influence of magnetic flux generation and reduce external noise (i.e., generated noise voltage causing mal-signaling or mal-triggering).

Caution is required with regard to the overall inductance and capacitance in the whole driver design layout as an LC resonance depending on the values of these components, including the parasitic values, may cause an unwanted oscillation during

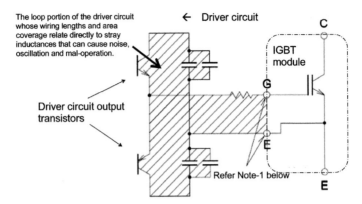

The loop portion of the driver circuit whose wiring lengths and area coverage relate directly to stray inductances that can cause noise, oscillation and mal-operation.

Driver circuit output transistors

Note 1: Use of +ve/-ve line parallel patterning directly on PCB should be first design choice for the connection here. In the case a wiring connection becomes essential, use of twisted +ve/-ve wire-pair or common-core shielded type wire is recommended.

Figure 3.67 Consideration for reducing gate wiring inductance in an IGBT module drive circuit design.

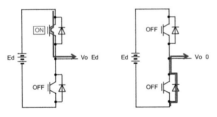

The mode when current flows out of an inverter phase:

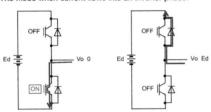

The mode when current flows into an inverter phase:

When an upper arm element is turned on, emitter of the lower arm element, dv / dt will take on each switching. Since the displacement current proportional to:

$$i = C_{CG} \frac{dE_d}{dt_{sw}}$$

at this time flows into the gate wiring, the feedback effect can cause the gate voltage of the turning OFF upper IGBT to swing back to a positive value, which can subsequently create mal-triggering. To suppress this effect, generally following measures are adopted in gate driver design process.

1) reduction of gate wiring impedance
2) application of a negative gate biasing (in the range of -5 to -15V) to keep the noise-effected floating gate voltage below $V_{GE(th)}$
3) Use of an external capacitor across each IGBT's G-E terminals to act as a filter against the abnormal voltage rise.

(Note: Further explanation of dv/dt coupling feedback mechanism is given in the following text and Figure 3.69.)

Figure 3.68 Maltriggering of the IGBT gate circuit caused by feedback from the device's capacitance.

operation, leading to device destruction. So proper analysis using simulation tools and experimental values is desired for establishing a good and rugged driver circuit design.

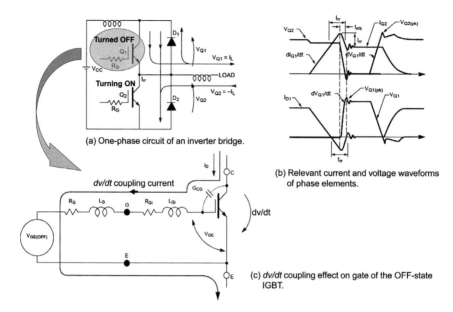

Figure 3.69 Mal-triggering caused by dv/dt coupling to the gate of IGBTs in hard-switching circuits.

The other important aspect in the IGBT module driver design, particularly for high-power inverter/chopper hard-switching applications, is to apply negative biasing for turn-off operation and off-state stability of the device gate, preventing its mal-triggering. The typical topology of the negative biasing scheme is shown in Fig. 3.67. The following describes a typical circuit function that can lead to a mal-triggering.

As shown in Fig. 3.69a,b, in half-bridge and inductive mode operation the IGBT that is in the off-state is subjected to a sharp rise of positive voltage due to the recovery process of its antiparallel diode. This dv/dt can be higher than the rate of rise of V_{CE} at turn-off of the IGBT. This dv/dt generates a current in the collector gate capacitance that flows into the gate drive circuit. (See Fig. 3.58). Although the gate is reverse-biased in the off-state, this current causes an increase of V_{GE} toward $V_{GE}(th)$ due to the gate circuit impedance. In the worst case, the threshold voltage is reached at the IGBT chip and turn-on of the IGBT is initiated, resulting in an arm shoot-through. The requirements to avoid this untimely turn-on are as follows:

1. V_G(off) should be sufficiently negative. (See Table 3.3.)
2. R_G in off-state should be low. (Recommended values are normally given in the device manufacturer's product specifications.)
3. Gate circuit inductance, L_G, should be minimized.

3.4.3.5 Power circuit impedance and overvoltage protection

In a typical hard-switching high-power bridge circuit topology, the impedance of the DC link power circuit (see Fig. 3.62 to Fig. 3.64), including the internal R-L-C components of the bulky link capacitors and that of the connected power modules, plays an important role in describing the transient response of the circuit and requires due consideration in designing overvoltage protection by methods such as integration of appropriate snubber components. The following description explains the issue involving a power circuit's transient response and considerations required to apply a snubber circuit to counter its adverse effect.

Figure 3.70 illustrates a simplified power circuit and related switching waveforms of any one of the applied IGBT arm-switches in the bridge, focused only on the device's turn-off transient for simplification and with the equations characterizing transient values of the concern here. The energy that causes transient voltages in IGBT power circuits is proportional to $L_S I^2/2$. Here, L_S is the parasitic bus inductance and I is the operating current. An important fact to remember is that this energy is proportional to the square of the operating current. Therefore, high-current devices will require much lower power circuit inductance. This presents a challenge to the IGBT circuit designer because the physical size and thermal requirements of these devices make longer power circuit connections necessary. With conventional bus work, these longer connections will cause more parasitic inductance, making the snubber design very difficult.

For IPMs, IGBT modules, and discrete elements, design considerations for snubbers are very similar. To suppress the surge voltage (ΔV), the most important point is to reduce the parasitic inductance

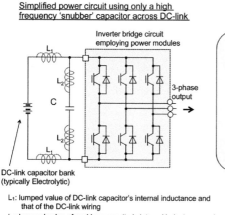
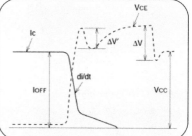

Figure 3.70 Simplified power circuit and waveforms for explaining snubber action for overvoltage suppression.

(L_1) of the main power circuit wiring. For using IGBT modules and discrete elements under such environments, application of driver circuit gate resistance higher than normal can be a partial solution for reducing the switching speed (i.e., lowering switching di/dt) of the device and, thereby, suppressing surge voltages during a dynamic switching operation. However, since the switching loss of the power device increases with the gate resistance, design optimization becomes a critical issue. Furthermore, due the existence of the snubber capacitor's lead inductance (L_2), the first peak surge voltage ($\Delta V'$) shown in Fig. 3.70 can become alarmingly high. For this reason due care is required in selecting snubber capacitors and their structure to match the to-be-protected device appropriately, keeping the leakage inductances as low as possible. Figure 3.71 provides samples of some actual snubber capacitor structures that have been designed to have very low lead inductance.

In the case of using a film capacitor in a snubber application, its preferable for its dielectric film inside to be of a polypropylene (PP) type, for its excellent frequency characteristic, instead of

The basic structure of the capacitor and the capacity

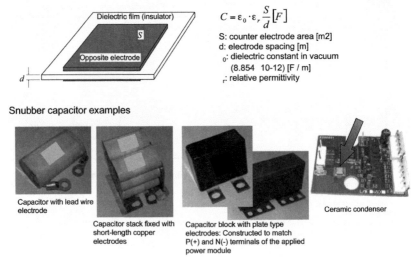

$$C = \varepsilon_0 \cdot \varepsilon_r \frac{S}{d} [F]$$

S: counter electrode area [m2]
d: electrode spacing [m]
ε_0: dielectric constant in vacuum
(8.854 10-12) [F / m]
ε_r: relative permittivity

Snubber capacitor examples

Capacitor with lead wire electrode

Capacitor stack fixed with short-length copper electrodes

Capacitor block with plate type electrodes: Constructed to match P(+) and N(-) terminals of the applied power module

Ceramic condenser

Figure 3.71 Basic structure and some practical examples of snubber capacitors.

a polyethylene terephthalate (PET), which is normally used as a capacitor across the AC line. Also, the internal inductance (equivalent series inductance [ESI], L) of a snubber capacitor should be as low as possible as its stored energy can become an unwanted cause of destruction of the power device by overvoltage generation rather than its protection. In recent years, application of a ceramic capacitor in a snubber circuit has become a choice for designers, particularly in the low-power-end (low-voltage) systems, due to its improved frequency and temperature characteristics. Ceramic capacitors have also become available in structures that are compact in size and are easy to mount on printed circuit boards (PCBs). These features simplify the applied system's assembly process and contribute in improving the applied snubber's performance by reduction of loop inductances.

Figure 3.72 shows four common snubber circuits applicable to power modules (IGBT and/or IPM) of different configurations (dual, six pack, seven pack, etc.). The A-type snubber circuit consists of a single low-inductance film capacitor connected from C1 to E2 on a dual IGBT module or from P to N on a six-pack module. In low-power

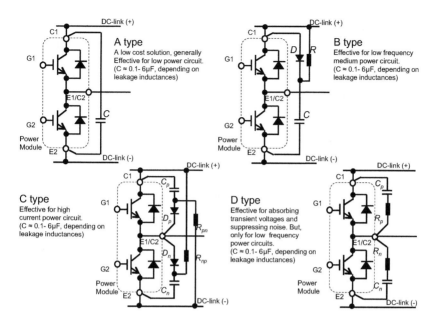

Figure 3.72 Different types of snubber configurations for power circuits.

designs this snubber will often provide effective, low-cost control of transient voltages. As power levels increase, an A-type snubber may begin to ring with parasitic bus inductance. The B-type snubber circuit solves this problem by using a fast-recovery diode to catch the transient voltage and block oscillations. The RC time constant of a B-type snubber should be approximately one-third of the switching period ($t = T/3 = 1/3f$). With large IGBTs operating at high-power levels, the parasitic loop inductance of the B-type snubber may become too high for it to effectively control transient voltages. In these high-current applications the C-type snubber circuit is usually used. This snubber functions similarly to the B-type, but it has lower loop inductance because it is connected directly to the collector and emitter of each IGBT of the bridge. The D-type snubber circuit is useful for controlling transient voltages, parasitic oscillations, and dv/dt noise. Unfortunately, its losses are quite high and it is generally not suitable for high-frequency applications. In very high power IGBT circuits, it is often helpful to use a small D-type snubber in conjunction with a main snubber of C-type in order to help control parasitic oscillations in the main snubber loop. In very-high-power applications it may be helpful to combine A-type and C-type in order to reduce the stresses on the snubber diode.

These snubber components and their values/characteristics are chosen in order to allow design of manageable snubbers while maintaining good control over transient voltages. In applications using modules of six-in-one or seven-in-one (six-pack or seven-pack) type, it is usually possible to use a single low-inductance capacitor connected across the P- and N-terminals as the A-type snubber shown in Fig. 3.72. Similarly, on dual-type modules a low-inductance capacitor connected between the C1 and E2 terminals is usually sufficient for control of transient voltages. Figure 3.72 also recommends the snubber capacitor value. The capacitor must be made large enough to avoid sympathetic oscillations in the inductive-capacitive (LC) circuit formed by the capacitor and the parasitic DC bus inductance. Usually a capacitance of about 1 µF per 100 A of collector current is sufficient. The exact type of capacitor and its value chosen in design may vary depending on power circuit architecture and application conditions, particularly related to leakage or stray inductances, associated with the main DC link bus design and the snubber circuit lead connections, operation frequency, and main power switching current value. The capacitor should be a PP film or a similar low-loss dielectric and be mounted as close to the module's terminals as possible. The total snubber loop inductance, including the capacitor's internal inductance, should be minimized (e.g., low-power circuit <10 nH, medium-power circuit <20 nH, and high-power circuit <30 nH). If parasitic oscillations are a problem in the application, it may be necessary to use a B-type snubber, as shown in Fig. 3.72. With high-current single-IGBT modules a single bus decoupling capacitor alone is usually insufficient for control of transient voltages. In these applications a clamp type R-C-D circuit, like the C-type snubber shown in Fig. 3.72, is usually used. In this circuit the snubber capacitors are charged to the DC bus voltage through the resistors. When the IGBT turns off, parasitic inductance in the DC bus causes a transient voltage across the IGBT. As soon as the voltage exceeds the DC bus voltage the snubber diode turns on and diverts the energy stored in the parasitic bus inductance into the snubber capacitor.

This snubber controls transient voltages better than any other type since it eliminates the inductance of the opposite IGBT package and the E1-to-C2 connection from the snubber loop. This clamp-type snubber circuit is typically constructed on a small PCB using axial or radial leaded capacitors along with fast-recovery snubber diodes and power resistors. The circuit board is then mounted to the bus bars directly above the power module. In this case the capacitor values are derived using equations given in Fig. 3.71 and assuming a transient voltage of 100 V with the IGBT switching at the rated current. To be effective the snubber must have a low inductance.

3.4.4 Power Circuit Thermal Design Aspects

When operating the power modules (IGBT and IPM) in inverter/chopper switching applications, the heat generated by their operation is mainly due to their conduction and switching power losses. The heat generated as a result of these losses must be dissipated away from the applied modules' internal power chips and into the environment using a heat sink. For these, thermal design of the whole system becomes very important. If an appropriate thermal system is not used the power devices will overheat, which could result in failure. Also, in many applications the maximum usable power output of the module will be limited by the system's thermal design. Furthermore, the critical delta-swing of temperatures (ΔT_j or ΔT_c, with reference to the ambient) may become worse due to improper thermal design and reduce endurance capability in terms of power cycling and thermal cycling and thereby the life of the applied device and consequently that of the whole system, as explained in Section 3.2.2. Therefore, appropriate estimation of operating power losses of the modules and an adequate thermal design, involving a heat sink of choice and optimum structure and an appropriate cooling method (free- or forced-air cooling, water or liquid cooling, etc.) to dissipate the heat aptly, are vital for system design and evaluation. The following provides guidelines to estimate operating power losses taking an IGBT module's application on a clamped-inductive-load inverter bridge topology as a typical example.

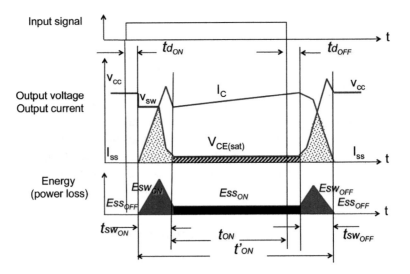

Figure 3.73 Waveforms showing the switching operation of an IGBT and its associated energies that relate to power losses in a typical clamped-inductive load inverter bridge arm-switch application.

3.4.4.1 Estimating power losses

The first step in thermal design is the estimation of total power loss. In power electronics circuits using IGBT modules the two most important sources of power dissipation that must be considered are conduction losses and switching losses. Figure 3.73 is an illustration of related voltage and current switching waveforms of an IGBT/diode arm-switch pair on an inverter bridge, similar to that described in Chapter 1. From such a clamped-inductive-load hard switching operation the two major power loss components can be estimated as follows:

Conduction losses: Conduction losses are the losses that occur while the IGBT is on and conducting current. The total power dissipation during conduction is computed by multiplying the on-state saturation voltage by the on-state current. In pulse width modulation (PWM) applications the conduction loss should be multiplied by the duty factor to obtain the average power dissipated.

A first approximation of conduction losses can be obtained by multiplying the IGBT's rated $V_{CE(sat)}$ by the expected average device current. In most applications the actual losses will be less because $V_{CE(sat)}$ is lower than the data sheet value at currents less than rated I_C. When switching inductive loads the conduction losses for the FWD must be considered. FWD losses can be approximated by multiplying the rated V_F given in the data sheet by the expected average diode current.

Switching losses: Switching loss is the power dissipated during the turn-on and turn-off switching transitions. In high-frequency PWM switching losses can be substantial and must be considered in thermal design.

The most accurate method for determining switching losses is to plot the I_C and V_{CE} waveforms during the switching transition. Multiply the waveforms point by point to get an instantaneous power waveform. The area under the power waveform is the switching energy expressed in watt-seconds/pulse or J/pulse. The standard definitions of turn-on ($\mathrm{Esw_{ON}}$) and turn-off ($\mathrm{Esw_{OFF}}$) switching energy are given in Fig. 3.73. The waveform shown is typical of the hard-switched clamped inductive load test that is used to generate all published switching energy data. The area is usually computed by graphic integration. Digital oscilloscopes with waveform-processing capability will greatly simplify switching loss calculations. From Fig. 3.73 it can be observed that there are pulses of power loss at turn-on and turn-off of the IGBT. The instantaneous junction temperature rise due to these pulses is not normally a concern because of their extremely short duration. However, the sum of these power losses in an application where the device is repetitively switching on and off can be significant.

In cases where the operating current and applied DC bus voltage are constant, $\mathrm{Esw_{ON}}$ and $\mathrm{Esw_{OFF}}$ are the same for every turn-on and turn-off event. The average switching power loss can be computed by taking the sum of $\mathrm{Esw_{ON}}$ and $\mathrm{Esw_{OFF}}$ and dividing by the switching period T. Note that dividing by the switching period is the same as multiplying by the frequency and results in

268 | *Applied Power Device Family*

the most basic equation for average switching power loss:

$$P_{SW} = f_{SW}(Esw_{ON} + Esw_{OFF}) \tag{3.14}$$

f_{SW} : switching frequency

$$Esw_{ON} = \int^{tsw_{ON}} I_C(t)V_{CE}(t)dt$$

: Turn-on switching energy illustrated in Fig. 3.73

$$Esw_{OFF} = \int^{tsw_{OFF}} I_C(t)V_{CE}(t)dt$$

: Turn-off switching energy illustrated in Fig. 3.73

Esw_{ON} and Esw_{OFF} can be found in the applied module's data sheets or specification. The turn-on loss includes the losses caused by the hard recovery of the opposite arm-switch's FWD in the phase of the inverter bridge. The critical conditions, including the junction temperature (T_j), DC bus voltage (V_{CC}), gate drive voltage V_{GE}, and series gate resistance (R_G), are usually given graphically in data sheets of the applied device.

In applications where the operating current and applied DC bus voltage are constant the average switching power loss can be computed by reading Esw_{ON} and Esw_{OFF} values from the data sheet curves at the operating current and using the equation given earlier. In applications where the current is changing, such as in a sinusoidal output inverter, the loss computation becomes more complex. In these cases it is necessary to consider the change in switching energy at each switching event over a fundamental cycle. A method for loss estimation in a sinusoidal output PWM inverter is given next. The final switching loss analysis should always be done with actual waveforms taken under worst-case operating conditions. The main use of the estimated power loss calculation is to provide a starting point for preliminary device selection. The final selection must be based on rigorous power and temperature rise calculations.

Loss calculation for sinusoidal VVVF inverter operation: One common application of power modules is the variable voltage variable frequency (VVVF) inverter. In VVVF inverters, PWM control is used to synthesize sinusoidal output currents. In this

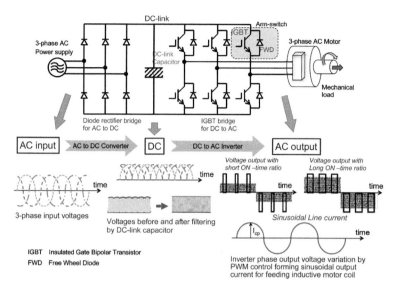

Figure 3.74 A typical VVVF inverter system and related waveforms of its operation.

application the IGBT current and duty cycle are constantly changing, making loss estimation very difficult. The following equations can be used for initial loss estimation in VVVF applications. Actual losses will depend on temperature, sinusoidal output frequency, output current ripple, and other factors. Figure 3.74 is a typical VVVF inverter circuit and output waveform.

IGBT power loss:

- Steady-state loss per switching IGBT.

$$P_{SS} = \frac{I_{CP} V_{CE}(\text{sat})}{2\pi} \int_0^\pi \frac{\sin 2x [1 + \sin(x+\theta)D]}{2} dx$$

$$= I_{CP} V_{CE}(\text{sat}) \left(\frac{1}{8} + \frac{D}{3\pi} \cos\theta \right) \quad (3.15)$$

$\cos\theta$: Power factor (pf) between inverter's output voltage and current.

D : Duty cycle of PWM control signal.

[Note: variation of PWM signal duty D in time t is given by; $\Delta D = (1 + D_{\text{sint}})/2$ and it corresponds to the inverter's output voltage change]

270 | Applied Power Device Family

- Switching loss per switching IGBT.

$$P_{SW} = \frac{(Esw_{ON} + Esw_{OFF})f_{SW}}{2\pi} \int_0^\pi \sin x \, dx$$

$$= \frac{(Esw_{ON} + Esw_{OFF})f_{SW}}{\pi} \tag{3.16}$$

- Total loss per IGBT.

$$P_Q = P_{SS} + P_{SW} \tag{3.17}$$

FWD power loss:

- Steady-state loss per FWD

$$P_{DC} = I_{EP} V_{EC} \left(\frac{1}{8} - \frac{D}{3\pi} \right) \cos\theta \tag{3.18}$$

- Recovery loss per FWD

$$P_{rr} = 0.125 I_{rr} t_{rr} V_{CE}(pk) f_{SW} \tag{3.19}$$

Power loss per arm-switch (shaded part of Fig. 3.74):

$$P_A = P_Q + P_D = P_{SS} + P_{SW} + P_{DC} + P_{rr} \tag{3.20}$$

Symbology:

Esw_{ON}: IGBT's turn-on switching energy per pulse at peak current I_{CP} and average T_j (e.g., $T_j = 125°C < T_{jmax}$ on data sheet)

Esw_{ON}: IGBT's turn-off switching energy per pulse at peak current I_{CP} and average T_j (e.g., $T_j = 125°C < T_{jmax}$ on data sheet)

f_{SW}: PWM switching frequency for every inverter arm-switch (normally, $f_{SW} =$ carrier frequency, f_C of the PWM control)

I_{CP}: Peak value of sinusoidal line output current of the inverter (Note: $I_{CP} = I_{EP} =$ corresponding arm-switch IGBT/FWD currents)

$V_{CE(sat)}$: IGBT saturation voltage drop at I_{CP} and average T_j (e.g., $T_j = 125°C < T_{jmax}$ on data sheet)

V_{EC}: FWD forward-voltage drop at I_{EP} and average T_j (e.g., $T_j = 125°C < T_{jmax}$ on data sheet)

D: PWM duty factor (modulation depth)

q: Phase angle between output voltage and current

I_{rr}: FWD peak recovery current (refer to Fig. 3.69)

t_{rr}: FWD reverse recovery time (refer to Fig. 3.69)

$V_{CE}(pk)$: Peak voltage across FWD at recovery ($V_{Q1}(pk)$ in Fig. 3.69)

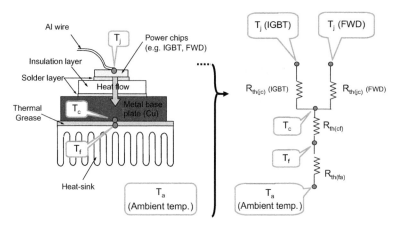

Figure 3.75 Thermal calculation model.

3.4.4.2 Estimating the junction temperature

The power chips in a power module (e.g., IGBT and FWD) have a maximum rated junction temperature defined by its manufacturer's data sheets or specification as the limiting condition for applying the module. This is normally 150°C or 175°C for silicon based devices. Good design practice is to limit the worst-case maximum junction temperature to 125°C or less. Reliability can be enhanced by operating the semiconductor junction at lower temperatures. If the total average power dissipated in the semiconductor device and the module base plate temperature are known, the junction temperature can be estimated using thermal resistance concepts shown in Fig. 3.75.

Thermal resistance (R_{th}) is specified on the selected power module's data sheets or specifications for use in thermal calculations. The junction temperature is estimated using the following equation:

$$T_j = T_C + P_T R_{th(jc)} \qquad (3.21)$$

$R_{th(jc)}$: Specified junction to case thermal resistance.

T_j : Semiconductor junction temperature.

$P_T = P_{SW} + P_{SS}$: Total average power dissipated in device.

T_C : Module's base plate temperature.

By using the appropriate values of $R_{th(jc)}$ and P_T the earlier equation can be used to estimate the junction temperature of either the IGBT or the FWD of any arm-switch of an inverter (refer to Fig. 3.74). For the initial design of heat sink systems, the contact thermal resistance is specified on the power module data sheets. The contact thermal resistance is the thermal resistance of the module–heat sink interface. The specified value assumes that a thermal interface compound such as white grease is used. The module base plate temperature can be estimated using the following equation:

$$T_C = T_a + P_T(R_{th(cf)} + R_{th(fa)}) \tag{3.22}$$

P_T : Total power dissipated by one arm-switch IGBT FWD pair (refer Fig. 3.74).

$R_{th(cf)}$: Thermal resistance of module and heat sink interface.

$R_{th(fa)}$: The heat sink to ambient thermal resistance.

T_a : Ambient temperature.

The value of $R_{th(cf)}$ is normally given for the entire module. For a reliable design, a thermal analysis is necessary using the measured base plate temperature and the total power loss under worst-case conditions.

Chapter 4

Future Prospects

A brief summary of achievements in the field of power semiconductor devices, starting from their bipolar-based origin to the latest IGBT modules and intelligent power modules (IPMs), has been made at the beginning of this chapter and is followed by an analysis of the changing requirements from the application fields and a projection of future growth for power semiconductors to comply with such needs. Such driving forces resulted in extensive research work to bring in use wide-bandgap (WBG) materials such as silicon carbide (SiC) and gallium nitride (GaN) for fabricating high-performance power devices in place of silicon (Si) and also to develop higher-functionality power modules based on and extending the original IPM technology platform. This chapter starts with a summary highlighting performance achievements of the silicon-based IGBT technology and the IPM family for power electronics applications. The chapter also includes discussions on future prospects related to the IGBT chip technology and more advanced fields, such as WBG-material-based power devices and trends toward higher levels of functionality integration over the IPM platform.

Power Devices for Efficient Energy Conversion
Gourab Majumdar and Ikunori Takata
Copyright © 2018 Pan Stanford Publishing Pte. Ltd.
ISBN 978-981-4774-18-5 (Hardcover), 978-1-351-26232-3 (eBook)
www.panstanford.com

4.1 Summarizing Device Achievements

Metal-oxide-semiconductor (MOS)-gated device concepts have played a primary role in driving the growth of power semiconductors since the late 1980s, thanks to the revolutionary achievements of insulated gate bipolar transistors (IGBTs) and intelligent power modules (IPMs) through several generations of refinements. The IPMs continued to offer an edge over equivalent IGBT modules in terms of performance, reliability, and system compactness by virtue of their integrated intelligence. In both cases of product evolution, refinement efforts were particularly aimed at improving performances of MOS-gated active switches relating to the reduction of voltage/resistance drop for a higher current conduction density or a lower on-state loss, energy/loss reduction for fast switching operation, and enhancement of the safe operating area (SOA) to withstand hard short-circuit stresses. Figure 4.1 explains the achievements with respect to breakdown voltage and

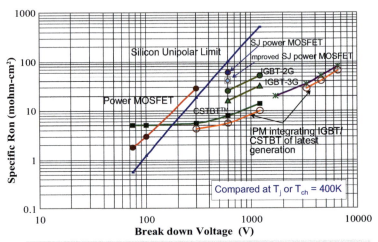

In terms of power losses, the users have benefited from continuous improvement made by various generations of IGBT families over the past 20 years

Note: The symbol "Ron" used here carries same meaning as the symbol used "r_{ON}" in the text elsewhere.

Figure 4.1 Static characteristics of Si MOSFET. Relationship between specific on-resistance and breakdown voltage.

specific on-resistance of power metal-oxide-semiconductor field-effect transistors (MOSFETs) and IGBTs in correlation with the theoretical silicon (Si) limit for unipolar structures, each assumed to be operating at a high chip temperature (400 K).

Although the IGBT is a bipolar device, the comparison in Fig. 4.1 is for the simple purpose of understanding an IGBT's performance improvement status in relation to the unipolar counterparts. In terms of power losses, users have benefited from continuous improvement made by various generations of IGBT families since its birth in the mid-1980s. Figure 4.1 also reveals that IPMs can offer better specific on-resistance than a plain IGBT device of the same generation, when compared at any particular breakdown rating. An IPM's advantage originates from the very concept of integrating a dedicated application-specific integration circuit (ASIC)-based drive and protection circuitry together with the principal IGBT power switch in smart module housing.

With the latest status of the IGBT being that it is the de facto power switch component for power electronics applications, the question arises as to the extent to which the IGBT's performance can be improved. Or what could possibly be the next device solution after the IGBT? Would it still be based on silicon? Or is it going to be realized by using a new material such as silicon carbide (SiC)? Although answers to these questions shall invariably depend on the market needs, a numerous number of researchers and engineers are engaged globally in advancing wide-bandgap (WBG)-material-based device solutions to a practical level as earliest as possible. The prospective candidates from the WBG group are considered to be SiC, gallium nitride (GaN), aluminum nitride (AlN), gallium oxide (Ga_2O_3), and diamond (C). Out of such possible postsilicon semiconductor materials for power devices, the most promising one till date is SiC, followed by GaN. In this chapter, application of the former for practical power module productization will be discussed in detail. Figure 4.2 is a review of Fig. 4.1 that includes comparisons of extrapolated theoretical and practical limits for silicon and SiC-based devices, indicating as well expectations from the use of SiC as the postsilicon material of choice.

Looking at Fig. 4.2, one can also infer that performances of various silicon-based devices such as IGBTs, power MOSFETs,

276 | Future Prospects

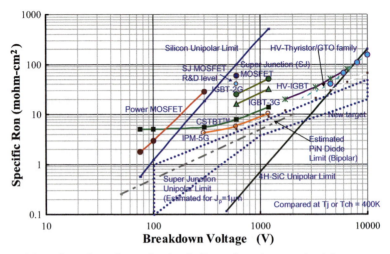

In terms of power losses, the users have benefited from continuous improvement made by various generations of IGBT families over the past 20 years

Note: The symbol "Ron" used here carries same meaning as the symbol used "r_{ON}" in the text elsewhere.

Figure 4.2 State-of-the-art static performances of various unipolar and bipolar devices in relation with simulated (theoretical) unipolar limits for Si- and SiC-based devices and a new target zone for the future devices.

superjunction (SJ) MOSFETs, p-n-junction-based diodes, and thyristor devices have improved close to their practical limits, even after incorporating the classic bipolar action and modulation effects as mechanisms for basic device function.

As said before, whatever be the choice of material to fabricate the next generation of power devices, the application needs that are faced today have to be squarely and aptly addressed. Figure 4.3 summarizes the major needs and technological hurdles that are to be challenged by new device solutions for the next generation of power conversion applications.

Along with such needs and hurdles, the other important direction that the power electronics equipment designs are expected to continue aspiring for is elevation of power density. As shown in Fig. 4.4, the power density has been remarkably improved in the past two decades by use of new power device components based on IGBT technologies, such as the different families and generations of IPMs.

Summarizing Device Achievements | 277

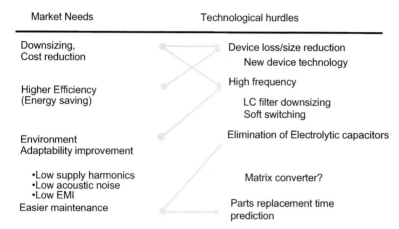

Figure 4.3 The major application needs and challenges for power devices.

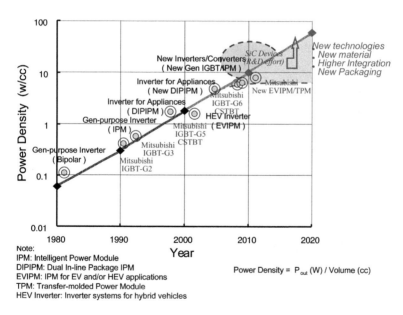

Note:
IPM: Intelligent Power Module
DIPIPM: Dual In-line Package IPM
EVIPM: IPM for EV and/or HEV applications
TPM: Transfer-molded Power Module
HEV Inverter: Inverter systems for hybrid vehicles

Power Density = P_{out} (W) / Volume (cc)

Figure 4.4 The trend of power density enhancement related to medium power class power electronics equipment.

Looking at the future, it appears that the introduction of new power conversion methods and/or highly integrated and intelligent, low-loss devices or component concepts will be very essential to raise the power density of equipment cost-effectively in 2010 and beyond, likely as extrapolated and indicated in Fig. 4.4. From the power device standpoint, possible silicon-based post-IGBT solutions are as important as those that can be realized by the prospective new semiconductor materials, such as silicon carbide. In the following sections, at first, the prospects of major silicon-based power semiconductor devices with their structural and operation principles will be discussed, followed by perspectives on realization of WBG-based devices in the future.

4.2 Future Prospects of Silicon-Based Power Device Technologies

As mentioned, this section gives an overview of key silicon-based devices, highlighting their achievements and their future prospects. The first part of Section 4.2.1 deals with the power MOSFET, as the user-friendly insulated gate feature of this transistor has continued to be the de facto standard in power switching applications. The second part deals with a more advanced power MOSFET, which is based on a unique SJ structure and exhibits improved performances. Sections 4.2.2 and 4.2.3, respectively, describe the statuses of two major silicon bipolar power device elements, the IGBT and the fast switching pin diode. These sections also provide perspectives on future prospects of these major devices.

4.2.1 Overview of the Power MOSFET and the Superjunction MOSFET

4.2.1.1 Power MOSFET

As explained before, power MOSFETs appeared in the late 1970s. Realization of this device became possible after MOS technology came into practice for microelectronic application. Unlike the bipolar junction transistor (BJT), the power MOSFET, which belongs to the field-effect transistor (FET) family, is a *unipolar* voltage-controlled device requiring very low driving power compared to the current-controlled BJT and conducts carriers only of the majority type in its on-state. Out of various categories of power MOSFETs in the FET family, the vertically structured n-channel enhancement mode type is the most popular one for power amplification and power conversion applications. Simplified basic forms of cell structures based on a planar-gate design and a trenched-gate design are shown in Fig. 4.5. This figure also illustrates the prime constituents of the device's drain-to-source on-resistance, and its specific value relates to the device's breakdown voltage in relation with the theoretical Si limit for a unipolar structure. As also described in the figure, a trenched-gate cell concept is very effective for reducing r_j and r_{CH} components of the total r_{ON}. As reveals, of the r_{ON} constituents described, r_{CH} becomes the most dominating factor for low-voltage class power MOSFETs, whereas r_D becomes dominating for the high-voltage range. By an optimized cell design, a trenched-gate structure significantly reduces the r_{CH} value as it nullifies the classic junction field-effect transistor (JFET) effect, which appears due to the accumulation of the majority carrier at the device's on-state in the upper-neck region below the gate overlapping area in a planer-gate structure (Fig. 4.5).

Therefore, refining of the basic trenched-gate cell has been the design trend, particularly applied to devices of the low-voltage (<200 V) class, and this trend is expected to continue in the future as well.

On the other hand, as the resistive component R_D increases exponentially with a higher breakdown voltage requirement, which in turn requires a thicker n-active drift layer, the power MOSFET has not been an appropriate solution for high-voltage applications. Thereby, its technological growth is mainly concentrated in the low-voltage (<200 V) fields. In such fields, advanced processes have been used to optimize the static on-state resistance versus breakdown voltage trade-off relationship. In recent years, the less than 0.2 μm design rule–based superfine trenched-gate power MOSFET process has been made practical for the less than 100 V class, where specific on-resistance (r_{ONsp}) values have reached very close to the theoretical Si limit [61].

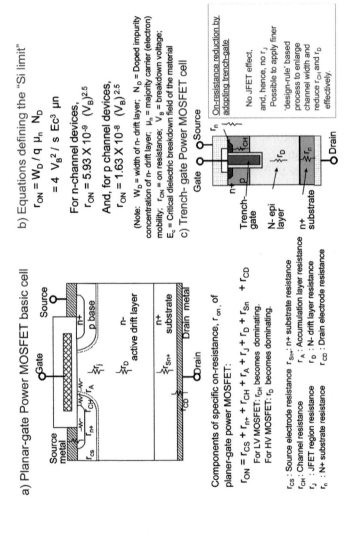

Figure 4.5 Power MOSFET structures, on-resistance r_{ON} components, and equations defining the theoretical Si limit for r_{ON} versus VB (breakdown voltage) trade-off.

Also, a power MOSFET can switch very fast because only the majority carriers contribute to its current flow, which eliminates any unwanted minority-carrier-decaying-phenomenon-related switching operation delay, which is typical of other bipolar devices.

Therefore, it has been the primary device solution for high-frequency applications. By comparison, bipolar transistors have greater power handling capability, but due to a slower switching speed characteristic, these devices have been mostly used for low-frequency type of applications. Whereas power MOSFETs have lesser power handling capability, the fast switching speed performance has made these particularly suitable for high-frequency applications.

4.2.1.2 Superjunction MOSFET

A pn SJ structure refers to a lateral formation of p- and n-type column pairs, effectively creating a much larger junction area than a simple 2D pn junction structure. Figure 4.6 gives a simplified comparison between the two structures and their operating principles.

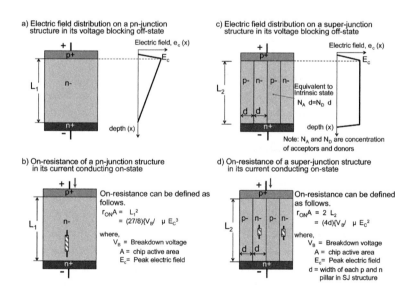

Figure 4.6 Comparison of structural features and the operating principle between a 2D p-n-junction and a superjunction.

One of the most important challenges pursued all through the evolution history of power semiconductors was to find methods to improve the trade-off between two key characteristics of any device: on-state voltage drop (on-resistance) and breakdown voltage. For power MOSFETS as well, this search has continued since its birth, generating many generations of device technologies until recently. More specifically, for power MOSFETs it translates into attaining the best trade-off between its specific on-resistance r_{ONsp} and breakdown voltage BV and to shrink the feature size without degrading device characteristics. However, for a simple 2D unipolar structure, like a planar or trenched-gate power MOSFET, a so-called silicon limit by device physics has been the maximum improvement that could be achieved. To achieve a breakthrough, a 3D structure based on the original SJ concept shown in Fig. 4.6 has been investigated since the late 1990s. The basic features of an SJ-concept-based 3D structure are explained in Fig. 4.6, in comparison with those of a plain p-n-junction-based one. Figure 4.6 also explains the differences in their operating principles. As for the electric field distribution under a voltage blocking state, a side-by-side stacking of p and n vertical columns allows lateral absorption and counterbalancing of excessive charges across each p and n vertical interfaces and, thus, a constant electric field distribution of a reduced magnitude can be achieved, even with a thin size and highly doped n columns. Thus, the specific on-resistance can also be reduced drastically as it is governed by two distinctly different physical equations, shown in Figs. 4.6b and 4.6c. This SJ principle has been extended to create practical MOS-gated power switching devices in the preceding decades. An SJ power MOSFET, which can also be said to operate under a voltage blocking state with a mechanism resembling a multi-RESURF-like operating principle, allows the doping level of its n-region to be typically 1 order of magnitude higher than that used in standard high-voltage MOSFETs. The additional charge generated due to such increased doping is counterbalanced by the adjacent charges of a p column, as also mentioned in Fig. 4.6. This counterbalancing, thus, contributes to a horizontal electrical field in the z axis direction (Fig. 4.6) without affecting the vertical field distribution. The electric field inside the structure is fixed by the net charge of the two oppositely doped columns. In this way, if both regions counterbalance each other perfectly, a nearly flat field distribution can be achieved, similar to that in the SJ p-n structure of Fig. 4.6. Practically the main advantage given by an SJ power MOSFET is a drastic reduction of the device area because of the device's low r_{sp} feature. Since the trade-off between r_{sp} and BV is dependent on the column pitch or width, refined process techniques for column fabrication are necessary for improving these characteristics. The comparably smaller chip size of an SJ-based power MOSFET, compared to a rating-equivalent conventional power MOSFET, leads also to a low gate charge characteristic, which, in turn, results in the former's higher-switching-frequency capability.

Some practical performances of SJ-concept-based power MOS-FETs are shown in Fig. 4.2 for comparison with other peers. As revealed by Fig. 4.2, the SJ concept, however, is not very practical for a voltage rating higher than 1000 V because of the difficulties in chip processing to achieve a high-quality charged-balanced structure having multiple pairs of vertically long p and n columns. Also, unlike a standard power MOSFET, the parasitically created integral body diode in an SJ structure has inferior reverse recovery and avalanche characteristics due to its multisectional and disintegrated space-charge balancing dynamic operation mechanism. Therefore, an SJ-based power MOSFET exhibits drawbacks in hard switching inductive load circuit applications, such as a motor-drive inverter system, where rugged switching operation of antiparallel diodes—primarily implying their ability to withstand high dv/dt stresses at reverse recovery turn-off switching cycles—is essential.

4.2.2 Review of IGBT Chip Technology and Its Future Prospects

The IGBT is now in widespread use among power modules for high-voltage and high-current applications because the device has well-balanced direct current (DC) and alternating current (AC) losses with an easy-to-drive feature by virtue of its MOS-gate structural feature. The easy-to-drive feature of the IGBT accelerated system integration in power modules. For example, to compensate for the weakness of the first-generation IGBT, a new device concept, named intelligent power module (IPM), was proposed by G. Majumdar of Mitsubishi Electric [53, 69, 83], the author of this book, opening the first era of system integration. The details of IPM's fundamental concept and its evolution can be found in the previous chapters of this book.

In 1968, Yamagami and Akagiri of Mitsubishi Electric introduced the basic idea of the IGBT as a pnp transistor driven by a MOSFET [22]. In the years between the late 1970s and the 1980s, J. D. Plummer and B. J. Baliga published their research work showing operation ability and features of a similar structure through papers and patents individually [55, 79, 80]. Though the most crucial barrier for practical use of the IGBT, particularly at its initial concept-building stage, was its inferior SOA due to the presence of its inherent parasitic thyristor element, discussed in Chapters 1 and 2 of this book, Nakagawa of Toshiba proposed an effective device structural solution for controlling the MOS channel current during device operation and called it the "non-latch-up IGBT" [60]. This was widely viewed as the cornerstone that gave birth to a usable IGBT power device for power conversion applications.

The development of the IGBT followed then after various R&D efforts and has fetched outstanding results in the preceding 25-year history of power devices. Many of the prior cell structures and process technologies contributed to the improvement of the IGBT. Since the IGBT has the structure of a double-diffused metal-oxide semiconductor (DMOS) transistor with a p^+ backside layer, almost all process steps of the IGBT are diverted from that of the DMOS transistor. The IGBT achieves a very high current density because of the conductivity modulation of electrons and holes, and the two-carrier current flow is optimized by a device simulator that was developed for analyzing thyristor action. Then, for improving the channel density for lowering on-resistance, the trenched-gated IGBT has been realized by using the leading-edge large-scale integration (LSI) process technology. After that, there were several important understandings and breakthroughs. In 1993, it was reported that suppressing the hole flow into the emitter electrode brings lower on-resistance by means of a higher carrier density around the upper surface. This technique came into wide use for IGBT development and is presently known as the injection enhancement (IE) effect [28].

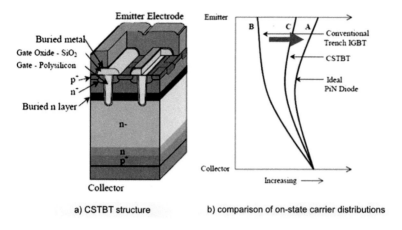

Figure 4.7 The novel structure and features of the CSTBT presented by Mitsubishi Electric in 1996. © [1996] IEEE. Reprinted with permission from Ref. [29].

A modified version of the IE effect–combined IGBT structure was proposed in 1993, called the carrier-stored trench-gate bipolar transistor (CSTBT) [29]. The CSTBT incorporated an additional thin n^+-layer right below the p-well on the emitter side of each trenched-gated transistor cell, as shown in Fig. 4.7.

Table 4.1 provides a comparison between an injection-enhanced insulated gate transistor (IEGT) and a CSTBT, highlighting their key features. Such concepts, together with thin-wafer-based light-punch-through-type vertical cell profile optimization, contributed immensely in expanding the limits of IGBT ratings—especially its voltage-handling capability—up to 6.5 kV level. An IGBT's wide SOA capability, compared to the power MOSFET and bipolar transistor, has also been theoretically explained and experimentally proven [26–28, 36–49, 59], and such valuable research works motivated device developers to develop very high current density type IGBTs of various voltage rating classes.

The advanced backside doping techniques on thin wafer proposed later are also worth mentioning as these concepts also have promising features [36, 62, 63]. Through such extensive R&D works on device structure so far, the power module based on IGBT chip performance has reached a rated current density four times

Table 4.1 Comparing key structural features of the IEGT and the CSTBT

Trench gate with wide cell pitch	
Gate design	

Structural features and key behavior	IEGT	CSTBT
	Non-channel region channel is floating	Non-channel region is connected to emitter
	Excessive electrons are injected from n^+ emitter	Carrier-stored (CS) layer n is at pn junction boundary

higher than the first-generation chips, introduced in the mid-1980s. Figure 4.5 gives an overview of the IGBT structural and performance evolution and key innovations made so far.

Figure 4.8 shows a summary of achievements so far in IGBT technology and its expected future trend as viewed by Mitsubishi Electric. The term "figure of merit" (FOM), expressed by the equation inserted in Fig. 4.9, relates to rated current density (J_C), on-state voltage (V_{on}) and turn off switching energy/loss (E_{off}) of a chip. This term was defined by the author and is generally used as an index validating the performance of a device of any generation applicable for power conversion.

4.2.3 Review of Diode Chip Technology and Its Future Prospects

In power conversion systems, silicon power diodes are widely used to perform functions such as rectification and freewheeling. All such applications demand that the operating diode's conducting on-state power loss is kept as low as possible.

In freewheeling applications where a diode is switched at a high frequency comparable to its complementary arm's transistor (e.g., an IGBT), its switching speed and switching power loss attributed by its reverse recovery process become additionally important. In such fast switching applications, a reverse recovery surge occurs

>> Introduction of trench-gate in combination with two novel structures aimed at refining carrier distribution in operation;
    - IEGT (IGBT with so called 'Injection Enhancement' effect),
    - CSTBT (IGBT with so called 'Carrier Storing' effect) ,
and thin wafer technology led to immense improvement of IGBT performance making it the core power switch of today.
>> Rated current density of the latest IGBT is four times higher than that of the 1st generation devices introduced in the mid-1980s.

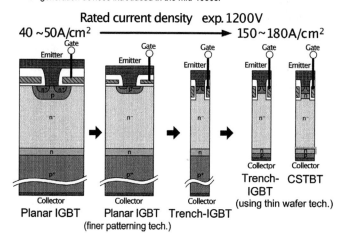

Figure 4.8 Structural changes of the IGBT cell design.

when a high-voltage power diode is switched from an on-state to an off-state. This surge can be the cause of a dielectric breakdown of the diode or its antiparallel IGBT on a typical inverter bridge circuit or can generate electromagnetic interference (EMI) noise. To solve such problems, the silicon high-voltage diodes, which are typically of the pin type in structural form for such applications (refer to the related description in previous chapters of this book), are required to have a soft recovery characteristic along with a low recovery loss performance. In other words, one of the main challenges in the design of a freewheeling diode (FWD) for inverter arm-switch application pairing with high-voltage/high-current IGBTs is to ensure a soft recovery behavior, especially under extreme operating conditions. This challenge comes along with the task of achieving low forward voltage drop and high switching speed characteristics for these diodes.

The reverse recovery surge depends on the variation of reverse recovery current with time (di/dt) and on parasitic inductance.

The di/dt increases with the amount of excessive minority carrier present in the drift region before the beginning of and during the reverse recovery process, as discussed in Chapter 3. The reverse recovery current can be "snappy" at the end section of the process or both voltage and current can be oscillatory caused by the inductive and capacitive components present in the circuit loop, added by the influence of the externally set switching speed (e.g., gate-driver conditions influence the turning-on speed of the diode's antiparallel IGBT), as also explained before. It has been explained that the enlargement of the depletion layer (space-charge) during the back end of the recovery process—and, hence, dv_R/dt of the reverse voltage build-up $V_R(t)$—across the diode should have to be controlled to avoid any quick removal of the remaining excessive carrier from the active region and thus avoid a destructive reverse voltage rise, that is, $V_R(pk) >$ breakdown voltage of the device. A lot of effort to achieve soft recovery has been made by optimizing the excess carrier distribution by low carrier injection and local carrier lifetime control, such as He^+ irradiation or Pt diffusion techniques, creating energy traps for recombining excess carrier and redistributing those in the active region of the diode. It has been explained in Chapter 3 that forced by circuit leakage impedance, L_{se} & R_{se}, the peak voltage at the back end of the reverse recovery process can reach a value determined by Eq. 4.1 given next.

$$V_{Rpk} = V_{CC} + L_{se}\frac{di_R}{dt} + R_{se}i_R(t)$$

$$=> \Delta V_R = V_{Rpk} - V_{CC} = L_{se}\frac{di_R}{dt} + R_{se}i_R(t) \qquad (4.1)$$

It has been also explained that the remaining excess carrier extraction may happen abruptly, causing snappy swings in the voltage $V_R(t)$ and current $i_R(t)$, leading to a damped $L_{eq}C_{eq}R_{eq}$ oscillation in the extreme case, where $L_{eq} = L_{se} + L_{si}$ (Fig. 3.60), C_{eq} is the equivalent circuit capacitance, and R_{eq} is the equivalent circuit resistance.

Satou et al. reported in 2000 that to suppress such oscillations in a high-voltage diode, the quality factor Q_f of the oscillating circuit should be as low as possible and to achieve this, they proposed a novel idea of creating a p^+-region at a selected location on the cathode side of the high-voltage silicon pin diode structure, as

Future Prospects of Silicon-Based Power Device Technologies | 289

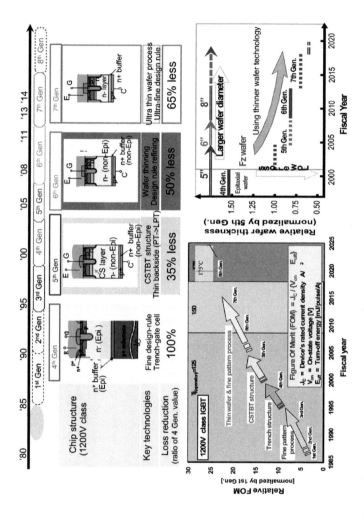

Figure 4.9 Key evolutionary achievements and trends in the development chronology of the silicon-IGBT chip technology (courtesy Mitsubishi Electric).

290 | Future Prospects

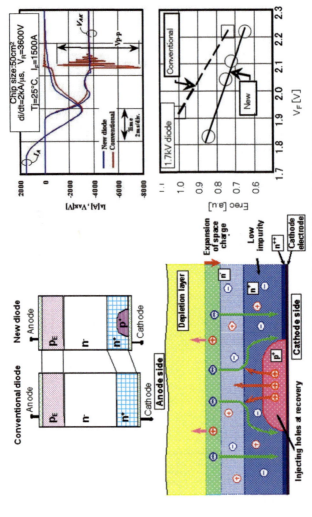

Figure 4.10 A novel diode design for oscillation-free operation and reduction of EMI noise (courtesy Mitsubishi Electric). © [2010] IEEE. Reprinted with permission from Ref. [84].

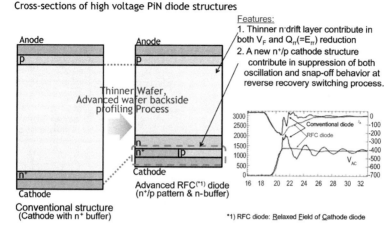

Figure 4.11 Advanced high-voltage pin diode, called the RFC diode, for fast switching freewheel operation using thin-wafer technology and backside profiling to embed a special n$^+$/p cathode structure (courtesy Mitsubishi Electric).

shown in Fig. 4.10. They explained that by optimizing the p$^+$-region pattern and doping, the recovery performance can be improved satisfactorily without sacrificing much on the device's on-state characteristic. In the 2010s, Mitsubishi introduced an advanced soft-switching high-voltage diode based on a modified new structure, named reduced field of cathode (RFC) technology, as shown in Fig. 4.11.

4.2.4 Integrated Power Chips Combining IGBT and Diode Functions

Two types of integrated power chip concepts combining IGBT and diode functions on a single chip emerged recently and have the potential to become the solution for many different power conversion applications.

One is called a reverse-conducting insulated gate bipolar transistor (RC-IGBT) and the other a reverse-blocking insulated gate bipolar transistor (RB-IGBT). The following section gives a brief explanation of the features of these integrated devices.

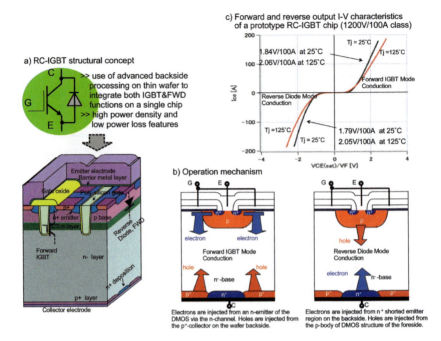

Figure 4.12 Reverse-conducting IGBT (RC-IGBT) structure, operation mechanism, and characteristics.

4.2.4.1 Reverse-conducting IGBT

In Fig. 4.12, the basic concept of structural integration to create a novel device called an RC-IGBT is explained [64].

The device shown here is based on the CSTBT structural design using a new thin-wafer technology. The important performances related to the forward and reverse conduction modes, along with prospects for adequate reverse recovery operation for the FWD of an experimental 1200 V/100 A chip having an active area of 1 cm^2 are also shown. Instead of forming the conventional stacking of an n-buffer layer and a p-collector layer, the striped n-region and striped p-region are independently formed on the wafer backside in optimized alignment with respect to the wafer's front-side trenched-gate stripes. In the case of the RC-IGBT concept as well, the wafer-thinning technology is expected to play again a vital role. Through various static and dynamic tests on prototype devices it has been found that this novel device has the potential to become a very attractive choice as a power switch for low-to-medium-power inverter applications in the near future.

The main idea behind the concept was to create a single-chip solution for an inverter arm-switch that can make systems compact and elevate their overall performance-to-cost ratio. The integration concept shown reveals that in an RC-IGBT, the built-in body diode acts as the device for reverse conduction, like a power MOSFET. However, this built-in body diode is created to have a low forward voltage as well as a fast switching recovery speed, matching high power hard switching inverter circuit requirements. To achieve an optimized integrated structure, advanced technologies, such as thin-wafer backside processing and selective helium irradiation of wafers for carrier lifetime control, are essential [64]. Experiments using He^+ irradiated on an RC-IGBT cell-based diode structure having a flat n^+ backside layer showed that the reverse recovery characteristic of the built-in body diode of an RC-IGBT can be tailored to the level of an equivalent fast-recovery-type FWD.

4.2.4.2 Reverse-blocking IGBT

In the early 2000s, a new type of AC–AC power conversion system called the AC matrix power converter started to receive attention as an alternative solution to AC–DC–AC voltage-fed power conversion systems (i.e., inverters) having a DC link storage base. Since its debut, the AC matrix power conversion technology has shown the potential to enter a variety of industrial applications that require to feed substantial energy back to the mains, such as in rolling mills, conveyor belts, and elevators.

The attractive features of the technology have, however, highly depended on the emergence of a cost-effective power semiconductor solution because the typical circuit topology requires a large number of devices to form the multiphase power conversion system.

Typically a matrix converter system requires a bidirectional power-switch component connected on each of its arms, or branches, with the capability of blocking voltage and conducting current in both directions, as illustrated in Fig. 4.13. In other

294 | *Future Prospects*

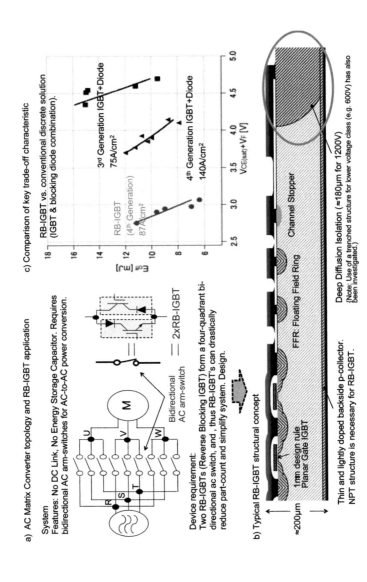

Figure 4.13 Topology, structure, and performance of reverse-blocking IGBT (RB-IGBT) for AC matrix converter application.

words, a four-quadrant bidirectional switch having fast switching and low-loss characteristics would be an ideal choice for a matrix converter to minimize its circuit cost. The initial solution to the system need was to use discrete power devices (e.g., IGBT and diode combination) in large numbers. One of the effective solutions proposed in the early 2000s is the RB-IGBT device concept, shown in the inset of Fig. 4.13 [65–67]. The key in this device structural concept is the use of a thin-wafer process technology and a deep-diffusion process to create an isolating wall at the edge of the chip structure. This has been applicable particularly for the voltage class up to a 1200 V rating. The alternative solution to the deep-diffusion-based isolation for a lower voltage range (e.g., 600 V rating) is proposed to be the fabrication of an oxide-filled vertical trenched wall [67]. Using such unique process techniques, RB-IGBT modules using high-current-type chips (e.g., 100 A rating) were reported and have been employed in actual systems.

Figure 4.13 also shows the most important characteristic trade-off of the new RB-IGBT chip-based module in comparison with an alternative choice of pairing discrete components (e.g., a discrete IGBT and a discrete fast switching diode). As is evident, the smartly integrated RB-IGBT concept, which is based on thin-wafer technology and special isolation creation processes such as the diffusion isolation (DI) technique or the trench isolation (TI) technique described in the same figure, Fig. 4.13, promises sufficient improvement in the application by drastically reducing the parts count and simplifying circuit design. However, the system's market acceptance and the unique process techniques to manufacture RB-IGBTs still appear to need further improvement to achieve appropriate cost-effectiveness.

4.3 Prospect of Using Wide-Bandgap Materials

In the preceding sections of this chapter, discussions have focused mainly on the prospects of silicon-based power semiconductors. However, silicon-based power switching devices are reaching fundamental limits imposed by the low breakdown field of the material, and a substantial improvement is considered to be achievable only by using a new semiconductor material with a wider bandgap and higher electrical field breakdown properties than silicon, such as silicon carbide. Material properties of SiC make it a natural choice for high-power, high-temperature, and high-frequency applications. The following section provides major features of some prospective WBG materials and some details about SiC-based devices, which are considered as the earliest feasible practical solutions.

4.3.1 WBG Material–Based Advanced Power Chip Technologies

After decades of research and commercialization of a multiple number of silicon-based power devices, such as IGBTs, MOSFETs, and IPMs, the so-called Si power is considered to be approaching its limit in terms of room for further performance improvement achievable for devices based on the material. For this reason, extensive research works are being carried out globally to bring in a new candidate material that can replace silicon. The interests have concentrated on materials having a physical bandgap property wider than Si's, or simply WBG materials, because this characteristic is essentially the key to improving performances of applied power semiconductors. Among several WBG materials tipped for power semiconductor application, SiC, GaN, Ga_2O_3, and diamond are noteworthy as each has attractive properties for becoming the base material for power semiconductors and has been investigated thoroughly in recent years. Research works have indicated that SiC of the 4H (hexagonal) crystal form type could be the most promising candidate for medium-to-high-power-range transistor and diode devices for its overall superiority, not only in terms of physical properties, but also in terms of feasibility of manufacturing with lesser complexity.

Its high electric breakdown property, of more than 3×10^6 V/cm, is 10 times higher than silicon's, allowing it to support a very high voltage across a thin layer. The high carrier drift velocity of over 2×10^7 cm/s avails a very fast removal of charge, which is essential for high-frequency operation, particularly for minority-carrier-driven bipolar devices. The high thermal conductivity, of

Table 4.2 Important physical properties of Si, GaN, and SiC

Material	Bandgap energy E_g eV	Dielectric constant ϵ_r (dimension)	Electron mobility μ cm²/V	Breakdown electric field EF 10^6 /cm	Saturated electron drift velocity V_s 10^7 cm/s	Thermal conductivity λ W/cm.K	FOM
Si	1.1	11.9	1500	0.3	1.0	1.5	1
GaN	3.4	9.5	900	2.6	2.5	1.3	407
3C-SiC	2.2	9.7	800	3.0	2.7	4.9	2381
4H-SiC	3.0	9.7	1000	3.5	2.7	4.9	3241
6H-SiC	2.9	9.7	460	3.0	2.0	4.9	1307

FOM $= \lambda \times$ Jonson FOM $= \lambda \times (E_c \times v_s)^2$. (Note: Also refer to Ref. [81].)

4.9 W/cmK, realizes high-temperature operation and better thermal management in power control applications.

Furthermore, the native oxide of SiC being SiO_2, the same oxide as silicon, the whole family of MOS-gated power devices used in silicon, that is, the power MOSFET and the IGBT, can all be fabricated in SiC. Because of superior material properties, SiC power devices can have specific on-resistance, up to several hundred times lower than similar devices in silicon. Silicon carbide exists in many crystalline forms, for example, cubic, hexagonal, and rhombohedral.

Table 4.2 summarizes some of the important physical properties of different polytypes of SiC in comparison with silicon and GaN, which is also being considered as a possible candidate for power semiconductor material in the future.

On the basis of the electric field profile within the drift region with a uniform doping concentration, the following equation shows that the specific on-resistance is related to the breakdown voltage:

$$r_{ON.sp \, (ideal)} = \frac{4(BV)^2}{\epsilon_r \mu E_c^3} \tag{4.2}$$

Here, BV is the breakdown voltage, ϵ_r is dielectric constant, μ is carrier mobility, E_c is critical breakdown electric field. On the basis of this assumption, various device structures were fabricated and evaluated in the past. The breakdown voltage versus the specific on-resistance graph shown in Fig. 4.14 maps results of many R&D attempts made in the early 2000s and performances of several practical n-channel SiC-MOSFETs presented by Mitsubishi Electric in the 2010s in comparison with the theoretical material limits of silicon and SiC-based unipolar devices derivable from the earlier

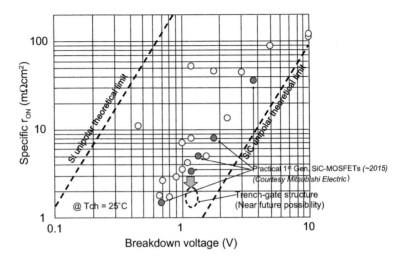

Figure 4.14 R&D efforts on improving the trade-off between the breakdown voltage and the on-resistance of SiC-MOSFETs.

equation. The results indicate that though remarkable improvement has been made recently in MOSFET structures, improving the BV versus r_{ONsp} trade-off, further efforts are necessary to get closer to the theoretical limits. The main difficulty in improving BV versus r_{ONsp} comes from the hurdle to achieving adequate carrier mobility in the channel region of a MOSFET structure due to surface irregularities across the channel region's interface with the gate-oxide layer, which effectively cause a large amount of trapped charge that does not contribute to conduction. This results in increased channel resistance, r_{CH}, of a planar MOSFET as explained in Fig. 4.5. The JFET region resistance, r_j (Fig. 4.5), also becomes a dominating item to affect the BV versus r_{ONsp} trade-off characteristic. New process techniques to enhance impurity doping in the JFET region and optimized postoxidation annealing under a specially controlled atmosphere have been effective in reducing r_{CH} and r_j considerably, thereby improving the BV versus r_{ONsp} trade-off of the latest power SiC-MOSFETs from 600 V to 3300 V voltage classes. In the years to come, channel mobility is expected to improve further by adoption of refined oxide formation processes or newer structures, such as the trenched-gate cell architecture shown in Fig. 4.5.

Through analyses to define an optimum power device solution using SiC, the unipolar MOSFET structure has proven to be the best choice for device implementation up to at least several thousand volts class ratings. Comparing total operation losses in the high-power three-phase inverter system, the SiC-device-based solution avails a large loss-saving advantage over the Si-IGBT-based solution, which can be 54% or larger if a carrier frequency of 14.5 kHz or higher is used. By experimentation it has also been found that the SiC-based MOSFET and Schottky barrier diode (SBD) combination can remarkably reduce power losses in actual system applications of a switching frequency higher than 20 kHz. Thus, SiC unipolar devices have a significant potential to replace bipolar IGBT devices in various application fields, including industrial and consumer motor drives, power supplies, automotive power electronics, and very high-voltage PE systems, in the near future. However, several critical issues still remain to be overcome to bring SiC devices in full-fledged commercial use. Some of these are listed next.

Key issues:

- Improving wafer size and quality
- Solving device processing hurdles
- Reducing wafer and device processing costs

Among these items, the most critical one is the availability of viable-quality low-cost SiC wafers from various sources in the global market. For a medium-power-range field, the power devices are expected to grow continuously, primarily depending on silicon-based chip and module technologies, for several more years and SiC is expected to become a material of choice for power semiconductors in the future. Figure 4.15 is a projection of major technological trends that are expected in the future.

4.3.2 Benefits from SiC Application

In general, SiC power semiconductor devices have the advantage of lower loss, even at very high frequency switching operation and they can work at a very high junction temperature condition.

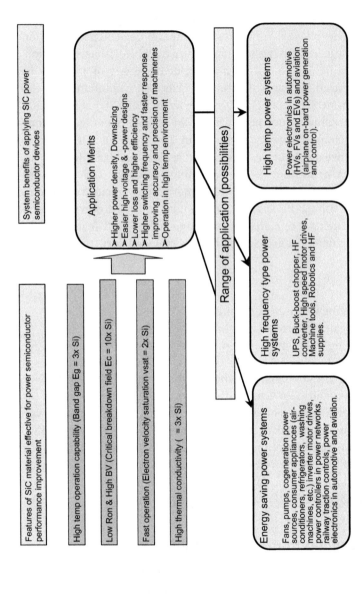

Figure 4.15 The important features of SiC material for power semiconductor applications and derivable benefits for system applications of SiC-based devices.

In addition, by using unipolar SBDs for freewheeling on a typical hard-switched inverter bridge, instead of pin diodes, the effects of the typical diode reverse recovery operation can be minimized, reducing EMI noise generation. In short SiC devices are expected to play a prime role in all system designs in the near future, where low conduction losses, low switching losses at high switching frequency of operation, and a high working temperature (above 200°C) would be essential for significantly advancing a system's performance. The predicted advantages of introducing SiC power semiconductor devices in various power electronics systems are summarized in Fig. 4.15 [61, 82].

The physical properties of the WBG and the high electric field breakdown explained before, in combination with other items considered in the FOM defined in Table 4.2, translate into improved efficiency, dynamic performance, and reliability of power electronics systems.

Thus, SiC power devices are expected to be used in various fields of power electronics equipment in the near future.

4.3.3 Status of SiC Devices

Many researchers have investigated various SiC devices, such as SBDs, pin rectifiers, power MOSFETs, JFETs, heterostructure field-effect transistors (HFETs), static induction devices, BJTs, and IGBTs. High-voltage capability has been demonstrated in almost all types of SiC power devices that have been experimented with under various research programs worldwide. The progress in terms of voltage-blocking capability has depended on the availability of low-doped thick epitaxial layers and improvements in the process steps and equipment for device fabrication. In the case of diode structures, significantly low on-resistance values have been achieved.

In terms of blocking voltage, SBDs have been fabricated having a breakdown voltage as high as 5 kV. Pin diodes have been experimented with for an even higher voltage range. However, pin diodes with blocking voltages above several thousand volts will require significant improvement of carrier lifetime in the low-doped region of the device structure. Also, with a thick low-doped drift layer for such high voltage structures using SiC, stacking faults originating from the so-called basal-plane dislocations are considered to cause a forward voltage drop drifting over time due to their recombination driven movements.

In the case of transistor-like controllable switching structures, JFET-type devices have exhibited fairly low on-resistance, even with a high-breakdown-voltage capability. However, a JFET is a normally on type of power switch and due to that functional feature, it faces many restrictions in power control circuit applications. For voltage classes up to about the 3 kV range, SiC-MOSFETs are expected to play a key role due to their normally off feature and simplicity in control requirement attributed to the virtues of their insulated gate structure, sufficient ruggedness in operation, and comparatively easier manufacturability. However, the on-resistance of the SiC-MOSFET experimented with has often shown to be much higher than the theoretical value predictable from SiC's physical properties. This is primarily attributed to a large amount of defect density generated at SiC/SiO_2 interfaces around the channel region of such device structures. The density of defect thus generated, which is also called the trap density, causes the carrier mobility to go down, decreasing the current conduction capability and increasing the on-resistance in device operation. Considering an n-channel SiC-MOSFET of 1200 V class, a channel mobility of ≥ 50 cm^2/Vs for electrons would be essential to reduce the on-resistance of the device to a sufficiently low value. Figure 4.16 shows a comparison of output characteristics, in terms of on-state voltage versus current density, of several state-of-the-art 1200 V class Si- and SiC-based transistors and diode devices, as reported by Mitsubishi Electric. From a simple perspective, SiC devices excel in performance at a low current density, whereas Si devices show higher levels of current saturation performance.

This feature allows design margin in terms of current rating and, hence, achievable power density to be higher, leading to a smaller chip and package in an application. The other noticeable feature is that the SiC device taken here being a unipolar MOSFET, the $V-I$ output characteristic starts from a zero-volt/zero-current origin, implying no built-in potential threshold in the device to initiate its current conduction, unlike a bipolar transistor such as an IGBT. The SiC-SBD, on the other hand, shows a built-

Prospect of Using Wide-Bandgap Materials | 303

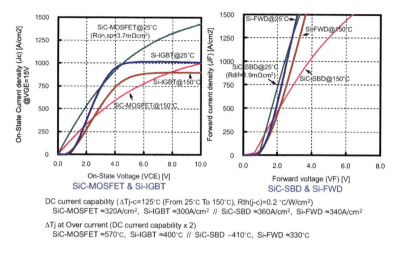

Figure 4.16 Comparison of output characteristics of state-of-the-art 1200 V class Si- and SiC-based transistors and diodes (courtesy Mitsubishi Electric).

in potential threshold. This property is a virtue of the barrier metal/semiconductor junction inherently existing in the device structure. This is in spite of the fact that the SiC-SBD's current is formed only by one carrier (electron) and, therefore, it is unipolar in category. The other noteworthy point is that the $V-I$ output characteristics of both the SiC-MOSFET and the SiC-SBD, unlike their silicon counterparts, show a positive temperature coefficient behavior, that is, the on-voltage increases with temperature rise for the entire range of conducting current. This behavior helps the chips of these devices to have a good current balancing feature when paralleled due to a positive feedback mechanism.

Figure 4.17 compares the inductive load turn-off switching operation of an IGBT and a SiC-MOSFET at identical turning-off current density and dv/dt. As revealed, the SiC-MOSFET turns off faster than an IGBT and is free of any tail current typically observed in the latter device. Figure 4.18 compares reverse recovery behaviors of a SiC-SBD and a Si-FWD at their turning-off operation on an inductive load circuit. As discussed before in this section, the SiC-SBD having a virtually zero level of stored minority carrier during its on-state, the reverse recovery charges are negligibly small. The small value of reverse recovery current visible in the

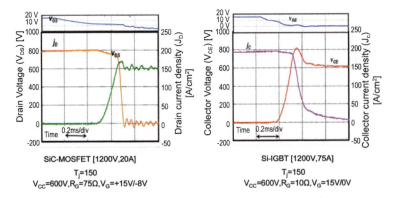

Figure 4.17 Switching waveforms comparing inductive load turn-off performances of a state-of-the-art 1200 V class Si-IGBT and SiC-MOSFET (courtesy Mitsubishi Electric).

waveform of Fig. 4.18 is a capacitive discharge current $\{= C(dv/dt)\}$ caused by the turn-off voltage transient applied to the device's bulk capacitance. Figure 4.18 shows temperature variations of on-state voltages at a high current density and unit values of turn-off switching energies in a mJ/A pulse for SiC-MOSFETs and Si-IGBTs when used in a typical inductive load circuit. The large change of on-state voltage of SiC-MOSFET over temperature is due its unipolar property discussed before. On the other hand, the turn-off switching energy (i.e., turn-off losses) of the same device does not change over temperature.

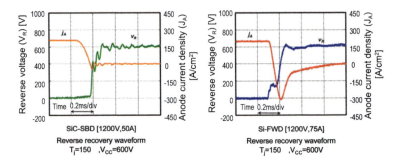

Figure 4.18 Switching waveforms comparing inductive load turn-off reverse recovery performances of state-of-the-art 1200 V class FWDs: Si-pin type versus SiC-SBD type (courtesy Mitsubishi Electric).

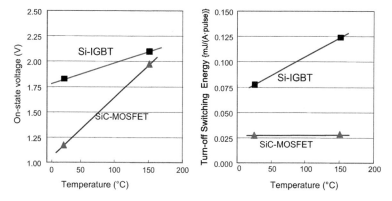

Figure 4.19 Temperature dependence of on-state voltage and turn-off switching loss characteristics of a state-of-the-art 1200 V Si-IGBT and SiC-MOSFET (courtesy Mitsubishi Electric).

Table 4.3 is a summary of the major characteristics of state-of-the art 1200 V class SiC and silicon-based transistors and diodes, the key pair of elements for power conversion electronics. By comparison, the SiC-MOSFET and SiC-SBD pair excels in dynamic switching performances, indicating their capabilities for high-frequency applications. However, the gate threshold voltage, $V_{GS(th)}$, of the SiC-MOSFET is quite low compared to that of the Si-IGBT ($V_{GE(th)}$). This requires high integrity in designing the gate driver circuit for the former transistor to enhance the circuit's immunity against noise and avoid mal-triggering. The other drawback of the SiC-MOSFET is its comparatively shorter time range (3 µs) capability to withstand a short-circuit condition, as indicated by short-circuit safe operating area (SCSOA) values in Table 4.3. The low gate threshold and narrower SCSOA for the SiC-MOSFET further indicate that a SiC-based IPM solution that can effectively counter these drawbacks is likely to provide better performance in power conversion applications than nonintelligent type of SiC-MOSFET modules, similar to advantages provided by silicon-based IPMs over IGBT modules.

Several aspects are explained earlier of state-of-the-art SiC-based power modules in comparison with their silicon-based counterparts, such as IGBT modules. Another competitive device

Table 4.3 Comparison of power-conversion-application-related major characteristics of state-of-the-art 1200 V class silicon- and SiC-based transistor and diode elements (courtesy Mitsubishi Electric)

	Unit	SiC-MOSFET	SiC-SBD	Si-IGBT	Si-FWD
Rated current density (J_C/J_D, J_F)	A/cm^2	320	360	300	340
On-state voltage ($J_{CE(sat)}$, $J_{DS(on)}$) @J_C/J_D, T_j=150°C	V	1.96		2.08	
Forward voltage (V_F) @J_F, T_j=150°C			1.73		1.83
Temp. coefficient of $V_{CE(sat)}$, $V_{DS(on)}$ @T_j= 25~150°C	mV/°C	6.5		2.1	
Temp. coefficient of V_F @T_j=150°C			2.1		1.1
Gate threshold voltage ($V_{GE(th)}$, $V_{GS(th)}$) @T_j= 25°C	V	2.0	—	6.5	—
[Inductive load] Turn-off switching energy (E_{Off}) @V_{CC}= 600V, T_j =150°C	mJ/ (A·pulse)	0.03		0.12	
Reverse recovery energy (E_{rr}) @V_{CC}= 600V, T_j=150°C			0.01		0.10
[Inductive load] Fall time (t_f) @V_{CC}= 600V, T_j= 25°C	μs	0.04		0.24	
Reverse recovery time (t_{rr}) @V_{CC}= 600V, T_j=150°C			0.03		0.17
Temp. coefficient of E_{OFF} @T_j= 25~150°C	μJ/ (A·pulse)	0.02		0.31	
Temp. coefficient of E_{rr} @T_j= 25~ 150°C			0.003		0.30
[Inductive load] SCSOA @V_{CC}= 800V, T_j=150°C	μs	3	—	5	—

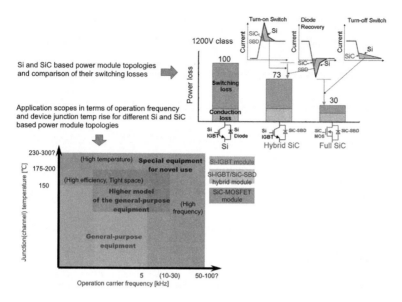

Figure 4.20 Preferable application scopes for different power module topologies using silicon and SiC devices.

solution for a wide range of power conversion applications is a hybrid combination of Si-IGBT and SiC-SBD power modules in which each integrated arm-switch is formed by antiparalleling a Si-IGBT for the transistor function and a SiC-SBD for the freewheeling function required for hard switching inverter/chopper bridge circuits. The inset in the top-right corner of Fig. 4.20 symbolically shows the connection topology of the hybrid SiC device in comparison with its full SiC-based and silicon-based counterparts and illustrates switching loss features of the three solutions. The turn-on loss of the arm-switch can be considerably low in the case of the hybrid combination because the applied diode is a unipolar SiC-based device having negligible reverse recovery charges, as explained earlier. Figure 4.20 also suggests preferable application ranges in terms of device junction temperature rise and its operating switching frequency for the three different module solutions combining silicon and SiC devices. In other words, the full-SiC-based module type is considered to suit the zones covering the high ends of device junction temperature and operating frequency,

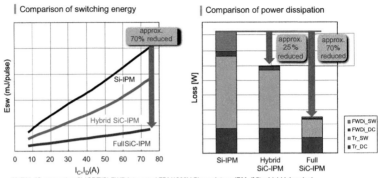

Figure 4.21 Simulated performances of 1200 V/75 A silicon- and SiC-based IPM devices (courtesy Mitsubishi Electric).

followed by the hybrid type and the silicon-based type, in turn. This proposition is also clear from the performances of three types shown by simulated plots given in Fig. 4.21. The modules chosen for comparison in Fig. 4.21 are all 75 A/1200 V rating and three-phase topologies introduced by Mitsubishi Electric. These modules are all IPM type, having dedicated driver and protection functions integrated. The advantages of applying SiC-based IPMs are discussed in the later part of this section. As revealed, a full-SiC module is estimated to reduce power losses in a 15 kHz carrier based three-phase pulse-width-modulated (PWM) inverter of 15 kW output by about 70% compared to an equivalent silicon-based device. On the other hand, by the same comparison, loss reduction achievable by a hybrid solution over the silicon-based one is approximately 25%. In case of the full-SiC-based IPM, both turn-on and turn-off switching losses, including the diode recovery switching, are very low compared to a Si-IGBT-based IPM.

This allows the device to operate at a carrier frequency of even a 15 kHz condition, as verified by the simulated results shown in Fig. 4.21. The low DC loss (or steady on-state power loss) for a SiC-IPM is a result contributed by the low on-resistance

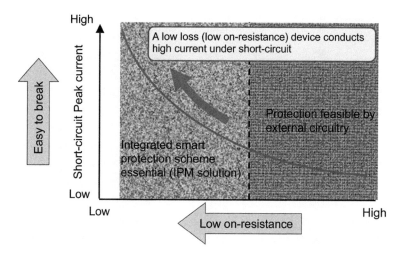

Figure 4.22 Application concerns over trade-off between short-circuit peak current and on-resistance for the SiC-MOSFET.

feature of the integrated 1200 V, 75 A SiC-MOSFET chip used in the module and is considered to be an essential feature in the application. Therefore, on-resistance reduction is a trend expected to continue in the evolution of SiC devices. However, this trend also increases concerns in application designs over the transistor's short-circuit-withstanding capability and, consequently, how to protect it appropriately under such abnormality without hampering the system's reliability. Figure 4.22 shows this behavior illustratively. As indicated, on-resistance of the SiC-MOSFET has an unavoidable trade-off relationship with short-circuit current, making a low-on-resistance type of device more vulnerable to destruction.

Under a low impedance dead-short condition at the output, terminals of a SiC-MOSFET-based inverter system, for example, would only have a few microseconds to protect the device and, hence, the system. To solve such an issue and fully utilize the advantage of a low on-resistance SiC-MOSFET for reducing power losses and increasing the power conversion efficiency of the applied system, the IPM-like concept of integrating a smart protection scheme within a power module appears very preferable, as indicated in Fig. 4.22.

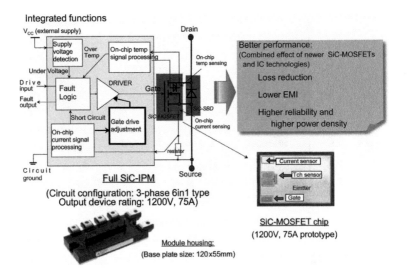

Figure 4.23 IPM-concept-based advanced SiC module comprising a 1200 V class highly functional SiC-MOSFET and SiC-SBD together with intelligent driver and protection schemes (courtesy Mitsubishi Electric).

Figure 4.23 shows a practical implementation of this idea. The full-SiC-based IPM using a 1200 V/75 A rated highly functional SiC-MOSFET and SiC-SBD chip pair to conform each arm-switch of a three-phase (six-pack) type of IPM is an advanced device solution for inverter applications introduced by Mitsubishi Electric in the early 2010s. The SiC-MOSFET power chip used for the system has on-chip current-sensing and temperature-sensing features. The feedback from these sensors is concurrently processed by the integrated logic and driving circuitry, and protection of the module against abnormalities such as short-circuit and overtemperature is internally performed. Short-circuit protection by this internal scheme typically takes 1–2 μs to shut down the faulted device and protect the whole module. Error signals are sent out for external processing. For protecting the SiC-MOSFETs against abnormal overtemperature, signals from on-chip temperature sensors are separately processed by a similar logic circuitry in the module. Together with other protection schemes, the SiC-IPM acts similarly to a Si-IGBT-based IPM, as discussed in the previous chapter. Such

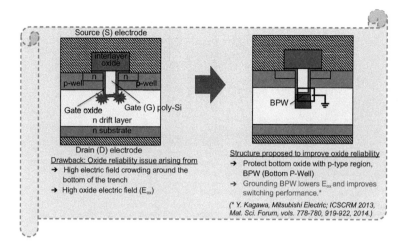

Figure 4.24 SiC-MOSFET evolution: planar-gate cell to trenched-gate cell with a special embedded layer.

an integration concept allows for use of low-on-resistance type of SiC transistors, like the SiC-based trenched-gate MOSFET shown in Fig. 4.24, and brings in higher efficiency benefit for power conversion systems.

Use of a trenched-gate structure for SiC-MOSFET cell design is considered to be very effective for on-resistance reduction and will most likely be a trend for designing devices covering a wide voltage rating range. However, one of the key concerns of using a trenched-gate cell structure has been crowding or intensifying the electric field, EF_{OX}, effect occurring at the trench bottom corners under a high-voltage-biasing condition, as shown in the illustration on the left in Fig. 4.24. This is alarming because the effect can lead to a gate-oxide breakdown phenomenon or a time-stressed degradation of the same, resulting in device destruction. To counter this effect a unique structural concept, shown in the illustration on the right in Fig. 4.24, was proposed by Mitsubishi Electric in 2013. The idea shown in the cell structure is to embed a p-well layer (bottom p-well [BPW]) at the bottom of the trenches and selectively connect or ground those to the source metal layer through filled contact holes internally. Creating embedded BPWs and connecting those selectively to the source metalizing layer (common circuit ground)

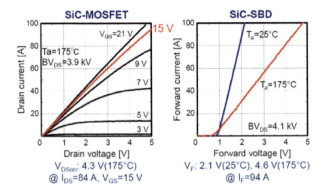

Figure 4.25 Characteristics and application features of the 3.3 kV class SiC-MOSFET and SiC-SBD chips for high-power-conversion applications (courtesy Mitsubishi Electric).

have been demonstrated to be done effectively by adding unique process steps in the device fabrication. The number of contact points and their layout distribution uniformly covering the whole chip active area are optimized to maintain a good balance between on-resistance reduction and short-circuit-withstanding characteristics.

SiC device development has also progressed to realize chip technologies for a voltage class higher than 1200 V. Figure 4.25 shows static output characteristics of a 3.3 kV rated chipset pair, SiC-MOSFET and SiC-SBD, developed for MVA-range high-power type of conversion systems, such as traction drives. Applying a high-current-rated 3.3 kV class full-SiC power module developed using these chips in 2–3 MW class high-power-level traction systems, large amounts of energy saving has been achieved, along with drastic reductions of both size and weight of the system.

Another example of a practical SiC module is shown in Fig. 4.26. A 75 A/600 V rated four-pack-type full-SiC-based IPM is shown to have been applied in 5.5 kW residential photovoltaic (PV) power conversion equipment, reducing power losses by almost 50% and

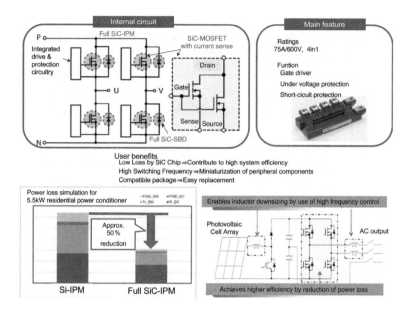

Figure 4.26 Characteristics and application features of a 600 V class full-SiC IPM for 5.5 kW residential PV power conversion applications (courtesy Mitsubishi Electric).

boosting the overall power conversion efficiency above 98%. In addition, by applying a high-frequency control scheme the size and weight of the system's filters have been reduced drastically. All these benefits have been the result of superior performances of the SiC-based power module.

4.3.4 Looking at Application Ranges for WBG Devices

Through our R&D efforts, it has also been demonstrated that an even larger scale of inverter size reduction and performance improvement can be achieved by using power devices based on silicon carbide and other WBG materials. Active development works are being carried out to generate future-generation power devices based on such materials to fulfill the future needs of power electronics systems.

Figure 4.27 projects footprints of power devices based on various semiconductor materials on an application mattress defined by an

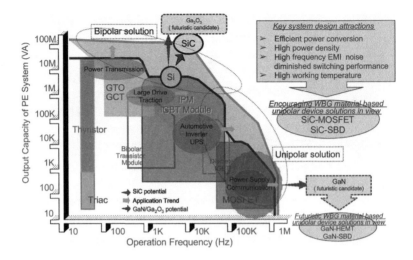

Figure 4.27 Possible enhancement of application range by WBG devices.

applied system's capacity versus operation frequency. The figure explains how silicon-based devices' boundary can be extended by devices based on WBG materials, such as SiC, GaN, and Ga_2O_3. Among these, prospects of greater application of SiC-based devices are considered to be more realistic currently than those based on the other two promising material candidates due to reasons of complexities in developing cost-effective fabrication technologies.

Finally, as the book has indicated all through, power electronics is the key technology field for efficient power conversion and power device technology is its most essential component field to support its growth. In the power electronics field, the inverter technology becomes the most effective reference as its core power circuit topology and function are often followed in many power electronics system designs. In this sense, the transitional steps that happened in the progress of inverter and its associated technologies are worth reviewing. Figure 5.26 summarizes such aspects.

As illustrated, the various power device technologies, such as the power module concept introduced in the mid-1970s; the epoch-making devices like the IGBT, introduced in the mid-1980s; the IPM, introduced in the late 1980s; and also the highly advanced versions of IGBT technologies, such the CSTBT, introduced in the late 1990s,

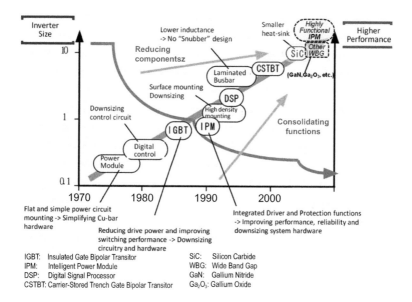

Figure 4.28 Transition of inverter technology as the barometer indicating growth of power electronics and key associated technologies, including power devices.

have contributed extraordinarily in downsizing and improving the performance of inverter systems. Through extensive global R&D efforts, it has also been demonstrated that an even larger scale of inverter size reduction and performance improvement can be achieved by using SiC-based power devices. The other promising materials for power devices are considered to be GaN and Ga_2O_3. And, in the far future, diamond (not shown in the figure) is also projected to be another candidate for power semiconductor applications, and elemental research works are also underway to investigate such possibility. Therefore, power electronics and power device technologies are expected to grow sustainably and provide support to society and counter climate change issues.

Bibliography

1. E. Ohno, *Introduction to Power Electronics*, Monographs in electrical and electronic engineering, Oxford Science, 1988.

2. W. E. Newell, Power electronics: emerging from limbo, *Power Electronics Specialists Conference (PESC) 1973 Record*, pp. 6–12, 1973.

3. H. Nishiumi, I. Takata, Y. Takagi, and S. Kojima, High voltage high power transistor modules for 440V line voltage applications, *Conference Record of IPEC-Tokyo*, pp. 297–305, 1983.

4. J. Nishizawa and Y. Watanabe, *Semiconductor Devices Including High Resistive Layer*, Japanese Patent No. 205068, Application date: Dec. 1950.

5. R. N. Hall and W. C. Dunlap, P-N junctions prepared by impurity diffusion, *Phys. Rev.*, **80**(3), pp. 467–468, 1950.

6. M. B. Prince, Diffused p-n junction silicon rectifiers, *Bell Syst. Tech. J.*, pp. 661–684, 1956, <http://www3.alcatel-lucent.com/bstj/vol35-1956/articles/bstj35-3-661.pdf>.

7. I. Takata, *Planar Semiconductor Device with Dual Conductivity Insulating Layers over Guard Rings*, US Patent No. US-4,691,224, Japanese Patent No. 1696252.

8. M. P. D. Burgess, *General Electric history*, <https://sites.google.com/site/transistorhistory/Home/us-semiconductor-manufacturers/general-electric-history>.

9. M. Mori, Y. Yasuda, N. Sakurai, and Y. Sugawara, A nobel soft and fast recovery diode (SFD) with thin p-layer formed by Al-Si electrode, *Proc. ISPSD 91*, pp. 113–117, 1991.

10. N. Soejima, *Diode Having Soft Recovery Characteristics over a Wide Range of Operating Conditions*, US Patent No. 5811873 A, 1995-05-18.

11. I. Takata, The latest IGBTs: from devices to modules, *Proc. Adv. Technol. Seminar 1996 IEEJ Annual Meeting*, pp. 17–25, 1996-3-27 (Japanese).

12. I. Takata, Problems on the SRH recombination model and a proposed solution, *Proc. IEEE*, pp. 193–196, 2006.

13. R. N. Hall, Germanium rectifier characteristics, *Phys. Rev.*, **83**, Session R12, p. 228, 1951.

14. W. Shockley and W. T. Read, Statistics of recombination of holes and electrons, *Phys. Rev.*, **87**(5), pp. 835–842, 1952.

15. I. Takata, On the diffusion component in leakage currents of silicon power devices, *Abstract of 2007 IEEJ Annual Meeting, 4-001*, **4**, pp. 1–2, 2007-03 (Japanese).

16. A. Herlet and K. Raithel, Forward characteristics of thyristors in the fired state, *Solid State Electron.*, **9**(11/12), pp. 1089–1105, 1966.

17. I. Takata, M. Bessho, K. Koyanagi, M. Akamatsu, K. Satou, K. Kurachi, and T. Nakagawa, Snubberless turn-off capability of four-inch 4.5kV GCT thyristor, *Proc. ISPSD 98*, pp. 177–200, 1998.

18. T. Sekiya, S. Furuhata, S. Itoh, and H. Haruki, FUJI high power transistors, *FUJI Electric Rev.*, **24**(2), pp. 55–59, 1978.

19. I. Takata, T. Hikichi, and M. Inoue, High voltage bipolar transistor with new concept, *Proc. IEEE*, pp. 1126–1134, 1992.

20. I. Takata, H. Nishiumi, H. Iwamoto, and Y. Yuu, A basic analysis of high voltage operation of high power transistors and diodes, *Conference Record of IEEE IAS 85*, pp. 900–904, 1985.

21. I. Takata, A consideration on the operating model of bipolar transistors, *2010 JEEE-document, EDD-10-111/ SPC-10-168*, pp. 63–68, 2010-11-30 (Japanese).

22. K. Yamagami and Y. Akagiri, *Transistor*, Japanese Patent No. S47-21739, File No. S43-87175.

23. I. Takata, A simple emulation of IGBT's forward action by a partial pin diode, *2009 IEEJ-document, EFM-09-025/EDD-09-059/SPC-09-126*, pp. 49–54, 2009-10-29 (Japanese).

24. T. Laska, M. Matschitsch, and W. Scholz, Ultrathin-wafer technology for a new 600V-NPT-IGBT, *Proc. ISPSD 97*, pp. 361–364, 1997.

25. A. Porst, Ultimate limits of an IGBT (MCT) for high voltage applications in conjunction with a diode, *Proc. ISPSD 94*, pp. 163–170, 1994.

26. H. Dettmer, W. Fichtner, F. Bauer, and T. Stockmeier, Punch-through IGBTs with homogeneous N-base operating at 4kV line voltage, *Proc. ISPSD 95*, pp. 492–496, 1995.

27. F. Bauer, H. Dettmer, W. Fichtner, H. Lendenmann, T. Stock Meier, and U. Thiemann, Design considerations and characteristics of rugged

punchthrough (PT) IGBTswith 4.5kV blocking capability, *Proc. ISPSD 96*, pp. 327–330, 1996.

28. M. Kitagawa, I. Omura, S. Hasegawa, T. Inoue, and A. Nakagawa, A 4500V injection enhanced insulated gate bipolar transistor (IEGT) operating in a mode similar to a thyristor, *Tech. Dig. IEDM 93*, pp. 679–682, 1993.

29. H. Takahashi, E. Haruguchi, H. Hagino, and T. Yamada, Carrier stored trench-gate bipolar transistor (CSTBT): a novel power device for high voltage application, *Proc. IEEE*, pp. 349–352, 1996.

30. M. Mori, Y. Uchino, J. Sakano, and H. Kobayashi, A novel high-conductivity IGBT (HiGT) with a short circuit capability, *Proc. ISPSD 98*, pp. 429–432, 1998.

31. T. Laska, M. Müunzer, F. Pfirsh, C. Schaeffer, and T. Schmidt, The field stop IGBT (FS IGBT): a new power device concept with a great improvement potential, *Proc. ISPSD 2000*, pp. 355–358, 2000.

32. M. Takei, T. Naito, and K. Ueno, The reverse blocking IGBT for matrix converterwith ultra-thinwafer technology, *Proc. ISPSD 03*, pp. 156–159, 2003.

33. G. Miller and J. Sack, A new concept for a non punch through IGBT with MOSFET like switching characteristics, *Proc. IEEE PESC 1989*, **I**, pp. 21–25, 1989.

34. M. Otsuki, Y. Onozawa, H. Kanemaru, Y. Seki, and T. Matsumoto, A study on the short-circuit capability of field-stop IGBTs, *Trans. IEEE ED*, **50**(6), pp. 1525–1531, 2003.

35. Z. Chen, K. Nakamura, and T. Terashima, LPT(II)-CSTBTTM (III) for high voltage application with ultra robust turn-off capability utilizing novel edge termination design, *Proc. ISPSD 12*, pp. 25–28, 2012.

36. K. Nakamura, K. Sadamatsu, D. Oya, H. Shigeoka, and K. Hatade, Wide cell pitch LPT(II)-CSTBTTM (III) technology rating up to 6500 V for low loss, *Proc. ISPSD 10*, pp. 387–390, 2010.

37. I. Takata, Destruction mechanism of PT and NPT-IGBTs in the short circuit operation: an estimation from the quasi-stationary simulations, *Proc. ISPSD 01*, pp. 327–330, 2001.

38. I. Takata, A trial simulation of the fourth secondary breakdown on IGBTs, *Proc. IPEC-Niigata*, S48-3, 2005.

39. H. Hagino, J. Yamashita, A. Uenishi, and H. Haruguchi, An experimental and numerical study on the forward biased SOA of IGBTs, *IEEE Trans. ED*, **43**(3), pp. 490–500, 1996.

40. M. Otsuki, Y. Onozawa, M. Kirisawa, H. Kanemaru, K. Yashihara, and Y. Seki, Investigation on the short-circuit capability of 1200V trench gate field-stop IGBTs, *Proc. ISPSD 02*, pp. 281–284, 2002.

41. J. Yamashita, H. Haruguchi, H. Takahashi, and H. Hagino, A study on the IGBTs turn-off failure and inhomogeneous operation, *Proc. IEEE*, pp. 45–50, 1994.

42. J. Yamashita, A. Uenishi, Y. Tomomatsu, H. Haruguchi, H. Takahashi, I. Takata, and H. Hagino, A study on the short circuit destruction of IGBTs, *Proc. IEEE*, pp. 35–40, 1993.

43. M. Yamaguchi, I. Omura, S. Urano, and T. Ogura, High-speed 600V NPTIGBT with unclamped inductive switching (UIS) capability, *Proc. ISPSD 03*, pp. 349–352, 2003.

44. E. Suekawa, Y. Tomomatsu, T. Enjoji, H. Kondoh, M. Takeda, and T. Yamada, High voltage IGBT (HV-IGBT) having $p + /p-$ collector region, *Proc. ISPSD 98*, pp. 249–252, 1998.

45. F. Bauer, N. Kaminski, S. Linder, and H. Zeller, A high voltage IGBT and diode chip set designed for the 2.8kV DC link level with short circuit capability extending to the maximum blocking voltage, *Proc. ISPSD 2000*, pp. 25–28, May 2000.

46. T. Wikström, F. Bauer, S. Linder, and W. Fichtner, Experimental study on plasma engineering in 6500V IGBTs, *Proc. ISPSD 2000*, pp. 37–40, 2000.

47. T. Fujii, K. Yoshikawa, T. Koga, A. Nishiura, Y. Takahashi, H. Kakiki, M. Ichijyou, and Y. Seki, 4.5kV-2000A power pack IGBT (ultra high power flat-packaged PT type RC-IGBT), *Proc. ISPSD 2000*, pp. 33–36, 2000.

48. K. Mochizuki, E. Suekawa, S. Iura, and K. Satoh, Development of 6.5kV class IGBT with wide safe operation area, *Proc. PCC 02*, **I**, pp. 248–252, 2002.

49. I. Takata, Non thermal destruction mechanisms of IGBTs in short circuit operation, *Proc. IEEE*, pp. 173–176, 2002.

50. I. F. Kovacevic, U. Drofenik, and J. W. Kolar, New physical model for lifetime estimation of power modules, *Proc. IPEC 2010*, pp. 2106–2114, 2010.

51. K. C. Norris and A. H. Landzberg, Reliability of controlled collapse interconnections, *IBM J. Res. Dev.*, **13**(3), pp. 266–271, 1969.

52. M. Ciappa, F. Carhognami, P. Cow, and W. Fitchner, Lifetime prediction and design reliability tests for power devices in automotive applications, *IEEE Trans. Device Mater. Reliab.*, **3**(4), pp. 191–196, 200.

53. G. Majumdar et al., A new series of smart controllers, *IEEE-IAS 1989 Proc.*, pp. 1356–1362.

54. G. Majumdar et al., Enhancing SOAs of IGBT modules for hard switching applications, *PCIM 1990 Proc.*

55. B. J. Baliga, M. S. Adler, P. V. Gray, R. P. Love, and N. Zommer, The insulated gate rectifier (IGR): a new power switching device, *IEDM 82 Tech. Dig.*, pp. 264–265, San Fransisco, USA, 1982.

56. M. Harada, T. Minato, H. Takahashi, H. Nishihara, K. Inoue, and I. Takata, 600V trench IGBT in comparison with planar IGBT: an evaluation of the limit of IGBT performance, *Proc. ISPSD 94*, pp. 411–416, 1994.

57. I. Takata, Fundamental difference between IGBT operation and bipolar junction transistor operation, *2008 IEEJ document, EDD-08-4 9/SPC-08-13*, 2008-10 (Japanese).

58. N. Tokura, Milestones achieved in IGBT development over the last 25 years (1984–2009), *IEEJ. D*, **131**(1), pp. 1–8, 2011-1 (Japanese).

59. R. Hotz, F. Bauer, and W. Fichtner, On-state and short circuit behavior of high voltage trench gate IGBTs in comparison with planar IGBTs, *Proc. ISPSD 95*, pp. 224–229.

60. A. Nakagawa, Non-latch-up 1200V 75A bipolar-mode MOSFET with large ASO, *IEEE Int. Electron Devices Meeting, Tech. Dig.*, pp. 860–861, 1984.

61. B. J. Baliga, Trends in power discrete devices, *Proc. ISPSD 98*, pp. 5–9.

62. T. Laska, M. Munzer, F. Pfirsch, C. Schaeffer, and T. Schmidt, The field stop IGBT (FS IGBT). A new power device concept with a great improvement potential, *Proc. ISPSD 2000*, pp. 355–358.

63. S. Dewar et al., Soft punch through (SPT) setting new standards in 1200V IGBT, *PCIM 2000 Proc.*

64. H. Takahashi, A. Yamamoto, S. Aono, and T. Minato, 1200V reverse conducting IGBT, *Proc. ISPSD 04*, pp. 133–136, 2004.

65. H. Takahashi, M. Kaneda, and T. Minato, Reverse blocking IGBT (RBIGBT) for AC matrix converter, *Proc. ISPSD 04*, pp. 121–124, 2004.

66. T. Naito et al., 1200V reverse blocking IGBT with low loss for matrix converter, *Proc. ISPSD 04*, pp. 125–128, 2004.

67. N. Tokuda et al., An ultra-small isolation area for 600V class reverse blocking IGBT with deep trench isolation process (TI-RB-IGBT), *Proc. ISPSD 04*, pp. 129–132, 2004.

68. K. H. Hussein, G. Majumdar et al., IPMs solving major reliability issues in automotive applications, *Proc. ISPSD 04*, pp. 89–92, 2004.

69. G. Majumdar et al., A new generation high performance intelligent power module, *PCIM 1992 Proc.*

70. H. Iwamoto, G. Majumdar et al., New intelligent power module for appliance motor control, *PCIM 2000 Proc.*

71. S. Shirakawa et al., A new version of transfer mold-type IPMs worth compact package, *PCIM China 2005 Proc.*

72. G. Majumdar et al., Miniature dual-in-line-package intelligent power modules, *Power Conversion 1999 Proc.*, pp. 93–100.

73. G. Majumdar et al., Compact IPMs transfer mold packages for low-power-motor drives, *Proc. ISPSD 04*, pp. 333–336, 2004.

74. T. Ueda et al., Simple, compact, robust and high-performance power module TPM (transfer-molded power module), *Proc. ISPSD 10*, pp. 47–50, 2010.

75. M. Ishihara et al., New transfer-mold power module series for automotive power-train-inverters, *PCIM 2012 Proc.*

76. M. Ishihara et al., New compact package power modules for electric and hybrid vehicles (J1-series), *PCIM 2014 Proc.*

77. S. Inokuchi et al., A new versatile intelligent power module (IPM) for EV and HEV applications, *PCIM 2014 Proc.*

78. K. H. Hussein et al., New compact, high performance 7th generation IGBT modulewith direct liquid cooling for EV/HEV inverters, *APEC 2015 Proc.*

79. B. W. Scharf and J. D. Plummer, A MOS-controlled triac device, *1978 IEEE Int. Solid-State Circuits Conf.*, Session XVI FAM 16.6.

80. B. J. Baliga, Enhancement and depletion mode vertical channel MOSgated thyristors, *Electron. Lett.*, **15**, p. 645, 1979.

81. C. M. Johnson, Comparison of silicon and silicon carbide semiconductors for a 10kV switching application, *IEEE-PESC Proc.*, pp. 572–578, 2004.

82. D. Stephani, Prospects of SiC power devices from the state of the art to future trends, *PCIM 2002 Proc.*

83. G. Majumdar, M. Fukunaga, and T. Ise, Trends of intelligent power module, *IEEJ Trans.*, **1**, pp. 1–11, 2007.

84. K. Satoh, K. Morishita, Y. Yamaguchi, N. Hirano, H. Iwamoto, and A. Kawakami, A newly structured high voltage diode highlighting oscillation free function in recovery process, *Proc. IEEE*, pp. 249–252, 2010.

Index

abnormalities, 96, 211, 232, 309–10

AC, *see* alternating current

activation energy, 103

alternating current (AC), 2, 8, 10, 21, 25, 30, 121, 127, 129, 146, 243, 255, 262, 283, 293

amplification, 17–18, 20, 23, 109, 129, 132, 220

amplifiers, 16, 18–19, 23, 140, 221

anode p-region, 71, 82, 84–85, 89, 161–62, 168

antiparallel IGBT, 287–88

application-specific integrated circuit (ASIC), 209, 215, 217, 221, 224, 231, 233–34, 275

arm-switch, 229, 231, 251, 268, 270, 272, 307, 310

ASIC, *see* application-specific integrated circuit

A-type snubber, 262–64

Auger recombination, 92, 98, 103

avalanche, 45–46, 60, 143, 174, 247

backside doping techniques, advanced, 285

backside profiling, 291

bandgap, 44, 296–97

base–collector portion, 62

base drive unit, 10, 123, 158

base electrodes, 9, 113, 123, 133, 174

base emitter, 81, 128

base plate, 193–95, 199, 201, 204, 221

base region, 109, 114, 125, 133–34, 136, 247

bevel-type edge termination, 182

biFET, 145, 158

bipolar devices, 32, 75, 93, 275–76, 281, 296

bipolar junction transistor (BJT), 18, 27–28, 30–32, 54–59, 83–85, 111–13, 117, 119–29, 131–36, 146–47, 157–59, 167, 173–75, 182, 278–79

bipolar transistor, 12–16, 28, 30, 80, 182, 184, 186, 213, 281, 285, 302

 carrier-stored trench-gate, 148, 285

 heterojunction, 16

 high-voltage insulated gate, 150

 injection-enhanced insulated gate, 148

 npn, 21

 pnp, 245

 real, 81

 reverse-blocking insulated gate, 291

BJT, *see* bipolar junction transistor

blocking voltages, 238, 293, 301

Boltzmann distribution law, 73–74

breakdown, 43, 45–46, 48–49, 51, 54, 58–59, 136, 179, 287, 311

Index

breakdown voltage, 43–44, 274, 276, 279–80, 282, 288, 297–98, 301

bridge circuit, 219, 237, 253, 287

B-type snubber, 263–64

capacitances, 18, 20, 30, 240–42, 244–46, 252, 257–58, 264, 304

capacitor, 10, 31, 47, 57, 68, 107, 140, 251, 253–55, 258, 260, 262, 264–65

carrier–carrier scattering, 81

carrier distributions, 75, 79, 105, 131, 151, 153, 155–56, 165

carrier mobility, 17, 297–98, 302

carrier-stored (CS), 148, 285–86, 315

carrier-stored trench-gate bipolar transistor (CSTBT), 148–49, 151–52, 154, 156, 165, 218, 224, 229, 231, 233, 274, 276, 285–86, 292, 314

case-type, 189–93, 197, 202–3

cathode n^+-region, 162

cathode n-region, 71, 165

central processing unit (CPU), 32, 139, 221

charge carriers, 73–74, 76, 80–81, 83, 95–96, 110, 128, 131, 133, 152, 165, 167–68, 177

charged carrier densities, 42, 45, 53, 82, 89–90, 94, 98–99, 102–3, 127–29, 152, 156, 165, 167–68

charged carriers, 40, 45–46, 74–75, 78, 85–86, 90, 95–96, 98–99, 101, 103, 105, 127, 159, 161, 167–68

charge neutrality, 39, 75, 77, 162

chip, 8, 62–63, 86, 136–39, 171–72, 181, 184, 195–96, 198, 207, 209, 217, 219–20, 223, 229, 283, 291–92, 302–3, 312

first-generation, 286

high-current-type, 295

integrated, 191

silicon-based, 299

chopper, 7, 12–13, 26, 63, 67, 70, 142–43, 187–88, 251, 259, 265, 300, 307

circuit, 18–23, 25–26, 58–59, 67, 142–44, 164, 182, 187–88, 195–96, 199, 233–35, 237–38, 249–60, 264–65, 288, 293, 295

driver, 242, 244, 254, 257

half-bridge, 245, 249, 251

main, 10, 30, 57–58, 107, 111, 242, 244

short, 12, 55–56, 232

coefficient of linear thermal expansion (CTE), 195, 201–2, 204

collector, 55–57, 113–14, 116–19, 121, 126–31, 133–34, 136, 145–46, 148–51, 156, 159–60, 164–67, 170, 237–38, 240–42

collector n^+-region, 126, 129–30

collector n-region, 240

collector p^+-region, 240

collector p-region, 146, 150, 156, 164, 168

COMFET, *see* conductivity modulated field-effect transistor

commutation, 30–31, 107, 250, 252

conduction band, 41, 96

conduction losses, 184, 237, 266–67, 301

conductivity modulated field-effect transistor (COMFET), 145

control circuits, 6, 253, 255

covalent bonds, 33–35, 38

CPU, *see* central processing unit

CS, *see* carrier-stored
CSTBT, *see* carrier-stored trench bipolar transistor
CTE, *see* coefficient of linear thermal expansion
C-type snubber, 263–64
currents, 46, 75, 105, 176–77, 234, 254, 267–68, 270

Darlington circuitry format, 182
Darlington connection, 119, 123, 146
  three-stage, 30, 157
  two-stage, 30
Darlington transistors, 8–9, 123, 133
DBC, *see* direct bonded copper
DC, *see* direct current
deep-diffusion process, 295
deformations, 200
  plastic, 200
  stress/strain, 199
degradation, 17, 180, 195
  time-stressed, 311
depletion layer, 66, 70, 74, 96, 98–100, 116, 128, 141, 145–46, 184, 288
destruction, 12, 54, 57–60, 71–72, 80, 85, 111, 121–22, 132, 134–35, 141, 143, 151, 171–75, 178–79
  #1-type, 173
  avalanche, 59, 143
  spontaneous, 132, 135–36
devices, 3–8, 12–13, 15–21, 27–32, 56–57, 59–61, 139–41, 214–15, 237–40, 245–47, 249–50, 281–86, 291–93, 301–4, 307–9
  advanced, 18
  alloy-treated, 61
  charge-controlled, 242, 244
  competitive, 305
  epoch-making, 314
  faulted, 310
  high-current, 260
  high-performance, 60
  high-power, 28, 200
  high-withstanding-voltage, 50
  integrated, 291
  large-area, 5
  lower-arm, 142
  low-loss, 278
  pin, 27
  pip, 109–10
  power-transistor-integrated, 187
  prototype, 292
  silicon-based, 275, 278, 308, 314
  thyristor-type, 31, 152
  variable-speed-type, 9
  voltage-controlled, 238, 279
device simulations, 159–60, 169, 171, 175
device simulator, 74, 88, 92, 103, 284
diffusion, 43–44, 51, 74, 79–80, 84, 86, 88–91, 99, 101–3, 118–19, 127, 137, 150, 295
diffusion current, 43, 51, 74, 79–80, 89, 99, 101–3, 127
diffusion self-alignment metal-oxide-semiconductor field-effect transistor (DSA-MOSFET), 137, 140, 144
diode, 4, 27–28, 49–50, 52–53, 59–65, 67–72, 77–78, 80–81, 83–85, 92–94, 105–7, 119, 164, 252, 286–88
  antiparallel, 188, 259, 283
  clamping, 188
  diffused, 62
  diffused junction, 61
  discrete fast switching, 295
  fast-recovery snubber, 265
  gold-diffused, 104

gold-diffusion, 105
hard-recovery, 71
high-speed switching, 66
high-voltage power, 287
p^+ in$^+$, 61
low-voltage, 65
p^+ nn$^+$, 48
partial, 164
platinum-diffused, 104–5
pn, 28, 59, 62
p-n-junction-based, 276
point contact, 60–61
rectification, 62, 66
reverse conducting, 112
snubber, 264
soft-and-fast-recovery, 85
DIPIPM, *see* dual-in-line intelligent
power module
direct bonded copper (DBC), 191,
193, 195–97, 199
direct current (DC), 2–4, 6–7, 21,
23, 25, 29–30, 106–7, 199,
250, 252, 264, 283, 293, 303,
308
discharge, 29, 68–69, 242, 246, 304
discharge tube, 2–3, 29, 46
DMOS, *see* double-diffused
metal-oxide-semiconductor
doping, 38, 43, 114, 154, 282, 291,
298
double-diffused metal-oxide-
semiconductor (DMOS), 14,
154, 284
drain, 18, 20, 65, 137, 139, 237–40,
243, 304, 311
drift, 39–40, 43, 51, 69, 74–75, 79,
82, 88, 111, 127, 167, 288, 297
drift current1 43, 51, 69, 79, 127
drift layer, 238, 241–42, 245, 247,
252, 279, 301, 311
drift mobilities, 51, 53, 66, 79,
82–83, 127
drift velocity, 62, 66, 127, 130, 132,
296

driving power, 12, 157–58, 213,
241, 244, 279
DSA-MOSFET, *see* diffusion
self-alignment metal-oxide-
semiconductor field-effect
transistor
dual-in-line intelligent power
module (DIPIPM), 13, 191,
208, 226–27, 255–56
dynamic withstanding voltage, 44

EB, *see* electron beam
Einstein's relation, 79, 127
electrical isolation, 191, 194, 198
electric charges, 29, 35, 42, 52, 64,
69, 73–74, 85, 145
electric field, 33, 35, 39, 42, 44–53,
73–74, 90, 110–11, 127–28,
130–32, 134, 166–68, 282,
297, 311
electric vehicle intelligent power
module (EV-IPM), 235
electric vehicles, 235
electrodes, 3, 41, 53, 60, 74–75,
78–80, 82, 99, 102, 124–25,
138–40, 168, 193–94, 196–97,
311
electromagnetic interference
(EMI), 63, 71, 85, 223–24,
252, 287
electron beam (EB), 66, 86, 91,
101, 147
electrons, 33–35, 39–41, 46, 53, 62,
79, 86, 88, 91, 159, 164, 238,
241, 247, 302–3
covalent, 34
covalent bond, 35
outermost shell, 35–37
secondary, 45
unpaired, 35
EMI, *see* electromagnetic
interference

Index | **327**

emitter, 113–14, 117–18, 121, 123, 125–29, 145–49, 152–57, 159–61, 165–68, 171–72, 174, 238, 240, 242, 284–86
energy, 1, 5, 21, 25, 36–37, 40–41, 44, 91, 179, 181, 200, 247, 260, 262, 264
energy gap, 36, 46, 77, 97, 100, 103
energy saving, 10, 14, 181–82, 185, 209, 277, 300, 312
epitaxial, 62, 116–19, 124, 145, 148–49
epoxy resin, 189, 197, 226
EPROM, *see* erasable programmable read-only memory
equivalent series inductance (ESI), 262
erasable programmable read-only memory (EPROM), 220
ESI, *see* equivalent series inductance
etching, 114, 137–38
EV-IPM, *see* electric vehicle intelligent power module

failure, 198–99, 202–3, 214, 248, 265
fatigue, 195, 200–201, 203
fault output (FO), 231–32
faults, 32, 106, 231–32, 302
FEA, *see* finite element analysis
Fermi energy, 40
Fermi level, 40–41, 96
FET, *see* field-effect transistor
field-effect transistor (FET), 18, 279
field stop insulated gate bipolar transistor (FS-IGBT), 148, 150–51, 172
figure of merit (FOM), 217, 227, 286, 297, 301
finite element analysis (FEA), 200

FO, *see* fault output
FOM, *see* figure of merit
force
counter electromotive, 58, 68–69, 136
induced electromagnetic, 22
reverse electromotive, 120
forward voltage, 62, 65, 74, 287, 293, 302–3, 306
free electron density, 38, 42, 44, 56, 68, 72, 75–79, 89, 96, 98, 101, 130–31, 163, 165
free electrons, 29, 35–42, 44–47, 51–52, 56–57, 64–70, 73–83, 87–90, 94–99, 101–2, 126–30, 132, 161–64, 167–68, 178–79
freewheeling, 251–52, 286, 301
freewheeling diode (FWD), 9, 58, 62, 64–66, 83–84, 142, 174, 196, 224, 229249–52, 267, 270–72, 287, 292–93, 308
frequency, 15, 18–21, 198–99, 243, 261–62, 267
audio, 114
carrier, 270, 299, 308
cut-off, 19–20
high, 277, 286, 299
high oscillating, 18
radio, 114
sinusoidal output, 269
variable voltage variable, 268
FS-IGBT, *see* field stop insulated gate bipolar transistor
functions, 15, 38, 106, 185, 188, 194, 211, 219–21, 223, 229, 276, 286, 314
coded-error-type FO, 231
data storing, 220
diagnostic, 215
exponential, 53
freewheeling, 307
high-frequency-amplification, 19
integrated, 187

intelligent peripheral circuit, 6
internal overtemperature
  protection, 223
inverter power conversion, 209
self-protection, 188
typical circuit, 259
unique signal transmission, 220
FWD, *see* freewheeling diode

Ga$_2$O$_3$, *see* gallium oxide
GaAs, *see* gallium arsenide
gallium arsenide (GaAs), 16, 28, 34
gallium nitride (GaN), 16, 34, 273,
  275, 296–97, 314–15
GaN, *see* gallium nitride
gallium oxide (Ga$_2$O$_3$), 275, 296,
  314–15
gate charge, 215, 225, 241–44, 283
gate circuit, 108, 215, 242–44, 255,
  259–60
gate commutated turn-off thyristor
  (GCT), 11, 13, 28, 31, 54,
  57–58, 83, 108–12, 156–57,
  167–68
gate controllability, 15, 22, 30, 214
gate drive, 111, 152, 175, 222, 224,
  231, 248, 251–55
gate drive circuit, 31, 109, 157,
  188, 241–42, 244, 256, 259
gate turn-off thyristor (GTO), 7–8,
  10–14, 27–28, 30–32, 54, 57,
  83, 107–9, 111–12, 120, 148,
  150, 152, 157, 167
gate voltage, 22, 52, 152, 169, 177,
  180, 215, 238, 240, 243–44,
  258
GCT, *see* gate commutated turn-off
  thyristor
germanium, 2, 4–5, 33, 46, 60–61,
  95, 103, 113–14
glow discharge, 46
gold diffusion, 84, 86, 91, 97
ground loops, 248, 253–54

GTO, *see* gate turn-off thyristor

hard switching, 26, 83, 283, 307
HBT, *see* heterojunction bipolar
  transistor
heat dissipation, 9, 186, 189,
  192–95, 223
heat sink, 10, 151, 192–93, 208,
  265, 271–72
HEMT, *see* high-electron-mobility
  transistor
heterojunction bipolar transistor
  (HBT), 16
heterostructure field-effect
  transistor (HFET), 28, 32, 52,
  59, 301
HEV-IPM, *see* hybrid electric
  vehicle intelligent power
  module
HFET, *see* heterostructure
  field-effect transistor
high-conductivity insulated gate
  transistor (HiGT), 148–49,
  152, 154–56
high-electron-mobility transistor
  (HEMT), 17–18, 28, 59
high-frequency MOSFET, 139
high-power Darlington-type npn
  bipolar transistor chip, 182
high-power IGBT circuits, 263
high-speed diode, 62, 68, 70, 80,
  83–86, 99, 142
high voltage, 8, 11, 30, 32, 43,
  45–46, 57–58, 61, 64, 83,
  116–17, 119–20, 134, 143,
  145–46
high-voltage BJT, 56, 83, 115, 122
high-voltage devices, 44, 50, 72,
  103, 122
high-voltage diodes, 53, 61, 63–64,
  66, 150, 287–88, 291

high-voltage integrated circuit (HVIC), 188, 209, 217, 219–20, 226
high-voltage intelligent power module (HV-IPM), 13, 235
high-voltage large-current device, 112
high-voltage operation, 32, 122, 129, 133–34, 168
high-voltage transistor, 8, 116–18
HiGT, *see* high-conductivity insulated gate transistor
hole density, 38, 40, 42, 68, 72–73, 75–80, 98, 101, 111, 165
holes, 33, 35–42, 44, 51–54, 64–70, 73–83, 86–90, 94–99, 101–2, 109–10, 112, 126–27, 132–34, 161–64, 167–68
HVIC, *see* high-voltage integrated circuit
HV-IPM, *see* high-voltage intelligent power module
hybrid electric vehicle intelligent power module (HEV-IPM), 235

ICs, *see* integrated circuits
IEC, *see* International Electrotechnical Commission
IEGT, *see* injection-enhanced insulated gate transistor
IGBT, *see* insulated gate bipolar transistor
  first-generation, 283
  high-speed, 174
  high-voltage, 111, 150, 154, 220
  high-voltage/high-current, 287
  n-channel-type, 21, 237
  non-latch-up, 284
  off-to-on, 252
  reverse blocking, 148, 293
  soft-punch-through, 151
  switching, 269–70

trenched-gated, 284
  uniform, 176
IGBT chip, 146, 151, 158, 171, 177, 194, 212, 221, 233–34, 245, 259, 273, 283, 285
IGBT devices, 215, 217, 237, 248, 275, 299
IGBT modules, 10–11, 146, 185–88, 191, 196, 211, 217–18, 221, 236, 245, 253–55, 258–61, 265–66, 274, 305
IGBT structure, 144, 152, 165, 170, 175, 213–14, 248, 285
IGR, *see* insulated gate rectifier
impact ionization, 44–46, 130, 132, 167–68, 179
impedance, 18, 31, 109, 111, 241, 244, 251, 254–55, 258, 260, 288, 309
impurities, 38–43, 46–47, 49, 52, 61–62, 77–78, 85–86, 96, 98, 113–17, 121, 125, 127–28, 141, 143
inductance, 30, 58, 191, 193, 246, 250, 254, 257, 261, 264–65
  internal, 262, 264
  lump, 249
  lumped bus, 251
  parasitic bus, 245, 249, 260, 263–64
inductive load, 21, 23, 25, 58, 245, 249, 251, 303–4, 306
infrared microscopy, 200
injection-enhanced insulated gate transistor (IEGT), 148–49, 151–54, 156, 165, 285–86
inner potential, 41–42, 74, 77
insulated gate bipolar transistor (IGBT) , 11–13, 30–32, 56–60, 67–71, 144–71, 153, 155, 157, 159, 161, 163, 165, 167, 169, 171, , 173–79, 211–19, 223–25, 240–42, 244–49,

257–60, 262–67, 269–72, 274–75, 283–86
insulated gate rectifier (IGR), 10, 145, 158
insulated gate transistor, 145, 148, 240
integrated circuits (ICs), 5, 8, 16, 32, 49, 136, 140, 158, 178, 209, 215, 217
intelligent functionality, 208, 223
intelligent functions, 189, 191
intelligent power module (IPM), 6, 11–13, 181, 184–89, 191, 208–9, 211–13, 215–27, 229, 231–33, 235–36, 273–76, 284, 308, 312–14
International Electrotechnical Commission (IEC), 196, 205
intrinsic carrier density, 38, 44, 46, 72
inverter, 9–12, 25–26, 62–63, 120–22, 124–25, 142, 158, 184, 187–88, 208–9, 255, 268–70, 293, 308, 314–15
  general-purpose, 9–10, 54, 83
  low-power, 256
  three-phase, 253–55
inverter bridge, 220, 236, 266, 268, 301
IPM, *see* intelligent power module
IPM series, 231–32
  advanced, 229
  fifth-generation, 223, 225
  first-generation, 225
  latest fifth-generation, 224
i-region, 41, 46–47, 65, 76–77, 79, 81, 82, 88–90, 93, 95, 97–99, 105, 117, 126–27, 141, 159, 168
irradiation, 44, 66, 86, 91, 100, 293
irregular pin diode, 78

JBS, *see* junction barrier Schottky

JFET, *see* junction field-effect transistor
junction, 62, 74–77, 79, 82, 90, 101–2, 114, 116, 168, 171, 238, 241, 271, 276, 281
junction barrier Schottky (JBS), 62
junction field-effect transistor (JFET), 16, 32, 52, 59, 279, 298, 301–2
junction temperature, 50, 198, 201–2, 267–68, 271–72

kinetic theory, 87

latch-up, 146–47, 158, 168–70, 214, 245, 248
lateral MOSFET (L-MOSFET), 144
law of mass action, 38, 73, 78–79, 90, 94
leakage, 22, 43–45, 50, 63–64, 67, 84, 86, 91, 97, 99–103, 171, 193, 264
leakage current, 91, 99–100, 105
leakage inductance, 248, 256–57, 261, 263
life endurance, 197, 200
life period, 197, 200–201
lifetime, 44, 68, 84–87, 90–92, 97–101, 103, 155, 174, 177, 197, 200, 242, 247, 301
lifetime control, 66, 84, 86–87, 90–92, 96, 112, 145–48, 150–51, 155, 166, 171, 288, 293
L-load, 30, 55, 83, 110, 120–22, 143, 170, 174–75, 177–79
L-MOSFET, *see* lateral MOSFET
load, 3, 25–26, 55, 57–58, 64, 174, 188, 225, 239, 242, 249–51
long-term reliability, 11, 50, 116, 122, 143, 148, 157
loop inductance, 253–55, 262–64

losses, 63, 107, 184, 263, 265–69, 283, 304
  electric, 107
  gate driver's, 111
  high-frequency, 195
  low, 63, 299
  low DC, 308
  switching transient-state, 192
  total operation, 299
low-voltage application-specific integrated circuits (LVICs), 217
low-voltage ASICs, 191, 215, 219
LVICs, see low-voltage application-specific integrated circuits

magnetic flux, 256–57
measuring circuit, 67
mercury rectifier, 3–4, 29, 46, 60–61
merged pin Schottky (MPS), 62–63, 84
MESFET, see metal-semiconductor field-effect transistor
metal-oxide semiconductor (MOS), 13, 137, 161–62, 184, 247, 274, 278, 284
metal-oxide-semiconductor field-effect transistor (MOSFET), 14, 19–20, 27–28, 30, 32, 56–57, 59, 136–37, 139–46, 141, 143, 157–59, 164, 186–87, 237–43, 274–76, 298
metal-semiconductor field-effect transistor (MESFET), 16
metal-semiconductor rectifier, 60, 62
minority carriers, 238, 240–42, 245, 247–48, 252, 288, 303
MMIC, see monolithic microwave integrated circuit

monolithic microwave integrated circuit (MMIC), 16
MOS, see metal-oxide semiconductor
MOSFET, see metal-oxide-semiconductor field-effect transistor
  normal, 141
  planar, 298
  unipolar, 302
MOS gates, 13–14, 137–39, 157, 186, 217, 220, 274, 283, 297
MPS, see merged pin Schottky

n-region, 39, 41–43, 46–47, 49, 52–53, 61–62, 65, 68, 70–77, 79–82, 85, 89, 95–97, 101–3, 107, 117, 137, 141, 149, 151, 154, 156, 157, 159, 162, 165, 168, 240–41, 282, 292
n^{-}-region, 130, 135, 150, 153
n^{+}-region, 39, 41–42, 48–49, 72, 81, 83, 126, 146, 161–62, 164
n-buffer, 145–46, 148, 150, 156, 164, 166, 286, 292
n-channel, 238–39, 241–44, 302
negative biasing, 255, 259
Newell, Dr. W. E., 5–7, 10–11, 120
noise, 17–18, 63, 223–25, 231, 248, 255–57, 263, 287, 301, 305
non-punch-through (NPT), 145–46, 148–50, 155–56, 160, 164, 166, 175
non-punch-through insulated gate bipolar transistor (NPT-IGBT), 148, 150, 155–56, 164, 166, 175
NPT, see non-punch-through
NPT-IGBT, see non-punch-through insulated gate bipolar transistor
n-region, 41–43, 46–47, 52–53, 61–62, 72–77, 79–82, 89,

95–97, 101–3, 141, 149, 154, 156–57, 159, 168
n-type semiconductors, 36–38, 64, 114, 238

off-/on-state, 65
off-operation, 143, 159, 170, 177–78
off-state, 22, 28–29, 151, 214, 239, 259–60, 287
Ohm's law, 43, 51, 237
on-chip, 221, 223–24, 229, 233, 310
on-off operation, 25, 192
on-operation, 30, 46, 70, 85, 97–101, 127, 143, 159, 162, 166–68, 174
on-resistance, 52, 142, 239, 274–75, 279–80, 282, 284, 297–98, 302, 308–9, 311–12
on-state, 21–22, 28–29, 31–32, 67–69, 71, 153, 156–57, 159, 215–17, 237, 239, 266, 279, 286–87, 302–6
on-voltage, 4, 12, 30, 61–66, 68, 84–86, 99, 111, 138–41, 145–47, 150–58, 166, 303
orbital, 34–36, 41
oscillations, 62, 71–72, 242, 244, 252, 257, 263–64, 288
OT, *see* overtemperature
overcurrent, 5, 12, 215, 221, 255
overtemperature (OT), 188, 221, 223–24, 232–33, 235, 310
overvoltage, 260–61

p^+-region, 49, 72, 83, 115, 125, 288, 291
pair generation, 36, 44, 51, 99–100
parallel IGBTs, 178

parasitic, 20, 23, 65, 122, 136, 142–43, 146, 211, 214, 238, 245, 248–49, 251, 263
parasitic inductance, 242, 244, 249–50, 260, 264, 287
parasitic oscillations, 263–64
parasitic thyristor, 168, 213–14, 221, 245, 248, 284
partial pin diode, 146, 156–57, 159–65, 167
PCB, *see* printed-circuit board
p-collector, 85, 148, 150, 286
PET, *see* polyethylene terephthalate
photovoltaic (PV), 209, 312
physical properties, 195–96, 296–97, 301–2
pin diode, 27–28, 46–48, 62, 64–72, 74–75, 78–81, 88–89, 92–95, 99–101, 103–5, 125–28, 141–43, 156–59, 167, 301
pin structure, 61–62, 112, 150
pip structure, 150, 168
planar structure, 11, 50, 63, 115, 117
pnp transistor, 105, 109, 113, 144–45, 158–59, 247, 284
Poisson equation, 87, 130, 134
polarity, 49, 136, 178, 199, 238, 250
polyethylene terephthalate (PET), 262
power amplification, 15, 17–18, 20, 279
power chips, 185–86, 191–93, 197, 199, 201–3, 208, 217, 223, 226, 229, 233, 265, 271
power circuit, 187, 248–50, 253, 255, 260, 263–65
power conversion, 1, 6, 11–12, 14, 21–23, 224, 276, 278–79, 284, 286, 291, 293, 305, 307, 311–13
power cycling, 200–203, 226, 265

power density, 212, 219, 276–78, 302
power devices, 6–8, 10–11, 14–15, 21–23, 27–28, 32–33, 43, 45–47, 51, 53–55, 59–62, 136–37, 184–86, 275–78, 313–15
power electronics, 1, 5–7, 9, 12, 14–15, 21–23, 106, 181–82, 184–85, 211–13, 273, 275–77, 299, 301, 313–15
power handling, 1, 7, 211, 227, 235, 281
power losses, 21–23, 25, 139, 142, 214–15, 217, 219, 221, 225, 227, 266–67, 270, 272, 274–76, 308–9
power modules, 9, 11, 184–86, 188–89, 191–93, 197, 199–205, 207, 209, 219–20, 235–37, 262–63, 265, 271–73, 283
power MOSFET, 14–15, 32, 137, 143–44, 213, 237–40, 242–45, 275, 278–79, 281–83, 285, 293, 297, 301
advanced, 278
n-channel, 21
planar-gate, 244
trench-gate, 282
power semiconductor devices, 2–3, 5, 14, 235, 273, 299, 301
power semiconductors, 6, 16, 21, 23, 181, 208, 217, 273–74, 282, 295–97, 299
power switches, 21, 23, 25, 216, 293, 302
power transistors, 5, 7–8, 10, 120, 157, 216, 233
p-region, 41–42, 46–47, 50, 52, 62, 72–77, 79, 81–82, 101–3, 105–6, 125, 141, 147, 154–57, 159–63

printed-circuit board (PCB), 16, 229, 262, 265
protection, 6, 12, 147, 186, 189, 191, 193, 209, 214–15, 221–23, 231–35, 254, 262, 308–10
short-circuit, 157, 232, 310
protection circuitry, 211, 231, 275
PT, *see* punch-through
p-type regions, 79, 238, 311
pulse width modulation (PWM), 266–70, 308
punch-through (PT), 66, 93, 104, 145–46, 150–51, 155, 166, 174, 271–72, 285
PV, *see* photovoltaic
p-well, 138, 142–43, 146–47, 149, 154–56, 160, 164, 167, 175, 177–78, 242, 285, 311
PWM, *see* pulse width modulation

quantum mechanics, 32, 34, 40

radar microwave, 61
RB-IGBT, *see* reverse-blocking insulated gate bipolar transistor
RC, *see* resistor capacitor
RC-IGBT, *see* reverse-conducting insulated gate bipolar transistor
reduced field of cathode (RFC), 291
reverse-blocking insulated gate bipolar transistor (RB-IGBT), 291, 294–95
reverse-conducting insulated gate bipolar transistor (RC-IGBT), 291–93
R&D, 226, 276, 284–85, 297–98, 313, 315

recombination, 35–36, 44, 51, 66, 80–81, 88–90, 92–99, 103, 105, 161, 247, 302
  direct, 93–96, 98, 100, 103–4
  linear, 95–96, 103–5
  radiative, 93–94
recombination velocity, 87–90, 96–97, 104
recovery, 59, 63–65, 67, 69–72, 84, 86, 90–91, 142, 251–52, 259, 270, 288, 291, 301
  fast switching, 293
  forward, 65–66
  hard, 70–71, 268
  large, 71
  snappy, 251–52
  soft, 71, 287–88
recovery destruction, 59, 67, 71–72
reliability, 14, 115, 143, 186, 197, 204–5, 208, 217, 221, 223, 235, 271, 274, 301, 309
resistance, 20, 29, 32, 44, 51, 59, 86, 116, 126, 140, 164, 169, 191–92, 195, 239–41
resistivity, 46, 48–50, 54, 61, 64–65, 116–17
resistor, 59, 61, 64, 107, 129, 255, 264
resistor capacitor (RC), 254
Restriction of Hazardous Substances (RoHS), 204
reverse recovery, 65–71, 83–85, 250–52, 283, 286–88, 292–93, 303–4, 306–7
reverse voltage, 29, 43, 46, 49, 141, 143, 288, 304
RFC, see reduced field of cathode
RMS, see root mean square
RoHS, see Restriction of Hazardous Substances
root mean square (RMS), 156, 199
ruggedness, 211, 245, 302

safe operating area (SOA), 22, 30, 51, 54–57, 60, 135, 170, 175, 214–16, 221, 274, 284–85, 305
saturation voltage drop, 184, 270
SBD, see Schottky barrier diode
SC, see short circuit
Schottky, 28, 61–63
Schottky barrier diode (SBD), 27, 32, 52, 62–65, 86, 142–43, 299, 301
SCR, see silicon-controlled rectifier
SCSOA, see short-circuit safe operating area
second breakdown, 7, 54–55, 57–59, 72, 120–21, 142, 179
selenium rectifier, 5, 60–61
semiconductor, 4, 6, 15–18, 29, 31–36, 38, 40–41, 43–44, 51–52, 62, 64, 78, 91, 93–94, 271
semiconductor devices, 1, 5, 21, 33, 36, 39, 41, 43, 45–47, 50, 55, 58, 60, 87, 98–99
  first, 106
  high-power switching, 106
  high-voltage, 63
  large-capacity, 72
  main high-power, 54
  silicon-based power, 278
  unipolar, 52
SFD, see soft-and-fast-recovery diode
Shockley–Read–Hall (SRH), 92, 99–100, 161, 166, 175
short circuit (SC), 12, 55–56, 215, 224, 232, 234
short-circuit condition, 55–57, 64, 83, 111, 121–22, 131–32, 135–36, 151, 166–67, 169, 171–73, 175, 214, 221
short-circuit safe operating area (SCSOA), 215, 305–6

SiC-based devices, 275–76, 296, 307, 314
SiC-IPM, 308, 310
SiC-MOSFET, 297–98, 302–5, 307–12
SiC-SBD, 303–5, 307, 310, 312
signal-to-noise ratio (S/N ratio), 17–18
Si-IGBT, 299, 303–5, 307–8, 310
silicon, 4, 6, 28–29, 36–38, 46, 50, 60–62, 64, 78–79, 181–82, 189, 271, 273–76, 296–97, 307–8
silicon-controlled rectifier (SCR), 1, 4, 12
silicon gel, 189, 195, 197, 225
sinusoidal, 268–70, 308
SJ, *see* superjunction
S/N ratio, *see* signal-to-noise ratio
snubber capacitor, 57, 107, 261–62, 264
snubber circuit, 26, 54, 107, 120, 260, 262, 264–65
snubbers, 10, 30, 57, 107–9, 120–21, 248, 260–61, 263–65
SOA, *see* safe operating area
soft-and-fast-recovery diode (SFD), 85
solder layer, 195, 200–202, 204, 271
source, 18, 20, 29, 44–45, 65, 113, 142–44, 153–54, 238, 240, 243, 246–47, 299–300, 311
space-charge, 283, 288
square SOA, double, 56, 175
SRH, *see* Shockley–Read–Hall
SRH-type recombination, 93, 95–100, 103–5
direct, 93–96, 98, 100, 103–4
linear, 95–96, 103–5
radiative, 93–94
static withstanding voltage, 44, 49–50

stray inductance, 10, 57–58, 67, 107–8, 139, 148, 174–75, 246, 254–55, 264
stresses, 57, 107, 119, 192, 200–202, 214, 264
cause-effect, 201
environmental, 189
hard short-circuit, 217, 274
longer-type periodic temperature swinging, 202
mechanical, 191
power cycling, 193, 202
short-circuit, 214
thermal, 192–93
thermal fatigue, 203
stress strain, 202
superjunction (SJ), 140–41, 140, 274, 276, 278, 281–83
surge, 10, 70–71, 250, 256, 287
surge voltage, 71, 107, 216, 249–51, 260–61
sustained operation, 176
switching circuit, 24, 65, 174, 242
switching cycles, 240, 242–43, 283
switching devices, 58, 63, 140, 143, 282, 295
switching energy, 22, 214, 217, 267–68, 270, 286, 304, 306, 308
switching frequency, 65, 84, 238, 241, 244, 268, 270, 299
switching losses, 22, 26, 63, 70, 150, 215–16, 223–25, 261, 266–68, 270, 301, 305, 308
switching operation, 23, 25, 63, 119, 217, 223, 237, 241, 244–45, 261, 266, 274, 283, 299, 303
switching power supply, 62–63, 65, 120, 140
switching waveforms, 22, 24, 260, 266, 304

thermal conductivity, 179, 191, 195, 204, 297

thermal cycling, 200–202, 226, 265

thermal design, 192, 265–67

thermal equilibrium, 38–42, 44, 73–74, 78–79, 94–95, 99–100, 102

thermal motion, 37–40, 44

thermal resistance, 172, 193, 203, 271–72

thermal runaway, 50, 59, 84, 142, 151, 172, 174, 203

threshold voltage, 241, 246, 259

thyristor, 4–7, 11–13, 27–28, 30–31, 50, 66–67, 83, 105–7, 109, 111–12, 120, 146–47, 167–70, 176, 178–81

thyristor Leonard system, 6

transient voltages, 249, 251, 260, 263–65

transistor module, 8–12, 28, 120, 122, 192

transistors, 2, 4, 7–11, 16, 18–21, 27–30, 54–55, 83, 112–20, 122, 129, 142–44, 182, 188, 305

trenched-gate cell, 279, 298, 311

trench-IGBT, 147, 151–53, 155–56, 159, 164, 180

trench-MOSFET, 138

triple diffusion, 118

U-MOSFET, 144

U-groove MOSFET, 144

UIS, *see* unclamped inductive switching

unclamped inductive switching (UIS), 143, 174, 178

undervoltage (UV), 224, 232

unipolar, 32, 52, 64, 80, 142, 274–76, 279, 282, 297, 299, 301–4, 307

UV, *see* undervoltage

vacuum tube, 3, 60, 112

valence band, 41, 96

variable voltage variable frequency (VVVF), 268–69

VD-MOSFET, 237

vertical MOSFET (V-MOSFET), 144

vertically diffused metal-oxide semiconductor (VMOS), 14

V-groove MOSFET, 137–38, 144

VMOS, *see* vertically diffused metal-oxide semiconductor

V-MOSFET, *see* vertical MOSFET

voltage, 21–22, 25–26, 29–30, 44–45, 47–55, 57–59, 61–62, 64–65, 116–19, 121–22, 154–56, 237–44, 249–50, 252–54, 282–83

VVVF, *see* variable voltage variable frequency

wafer, 52, 84–86, 111, 115–16, 118, 120, 141, 145–48, 150–51, 285, 292–93, 299

waveforms, 21, 25, 59, 67–68, 70, 108–10, 134, 136, 171, 175–78, 241–42, 244, 256, 261, 266–69

WBG, *see* wide bandgap

wide bandgap (WBG), 44, 64, 273, 275, 278, 296, 301, 313–14

wireless local area network (WLAN), 18

WLAN, *see* wireless local area network

Zener diodes, 251

PGMO 06/01/2018